Digital Services in the 21st Century

Digital Services in the 21st Century
A Strategic and Business Perspective

Antonio Sánchez
Belén Carro
Universidad de Valladolid, Spain

The ComSoc Guides to Communications Technologies
Nim K. Cheung, *Series Editor*
Richard Lau, *Associate Series Editor*

IEEE PRESS

WILEY

Published by John Wiley & Sons, Inc., Hoboken, New Jersey.
Published simultaneously in Canada.

For general information on our other products and services or for technical support, please contact our Customer Care Department within the United States at (800) 762-2974, outside the United States at (317) 572-3993 or fax (317) 572-4002.

Wiley also publishes its books in a variety of electronic formats. Some content that appears in print may not be available in electronic formats. For more information about Wiley products, visit our web site at www.wiley.com.

Library of Congress Cataloging-in-Publication Data is available.

ISBN: 978-1-119-31485-1

Printed in the United States of America.

10 9 8 7 6 5 4 3 2 1

To Diego

Contents

Contents   ix

Foreword

If there had been any doubt that our global society has moved into a digital century, it has surely been dispelled by the dramatic adoption of mobile phones now estimated to reach 4.6 billion subscribers, some of whom having more than one mobile. As to the so-called "smartphones," there are estimated to be about 3 billion subscribers to these devices. These numbers will almost certainly increase as the "Internet of Things" evolves and the associated devices begin to populate residential and enterprise operations at paces comparable to the mobile phone influx of the past decade. Already very dependent on devices linked to the Internet and the services they offer or rely upon, it seems predictable that our dependency will grow and that our vulnerability to buggy software or malware will increase.

There are many implications of this observation. The first is that we will need to prepare for a fragile future and that we need the makers of these devices and their software to be extremely thoughtful and careful to avoid major vulnerabilities. Second, we will need to update software in these myriad devices in a way that is secure and ensures with high probability that a valid update has been recorded. Third, we will be faced with configuring increasing numbers of devices and we need to make this process as simple, painless, and reliable as possible, especially as they change hands with people moving from place to place and inheriting or taking their many devices with them. Fourth, we will need to stay highly cognizant of the potential hazards to privacy that these programmable devices pose. Even simple information such as room temperature taken every few minutes might be useful for ascertaining whether people are present and where they are. At the least, a record of temperatures could reveal how many family members there are, what rooms they prefer, and when they might be away. Many other scenarios can be fashioned that highlight these risks and remind us that not all technology results in positive outcomes.

At the same time, we would be remiss not to speculate about the rich potential of artificial intelligence to smooth our interactions with a world populated with software-filled cyber-physical systems. Incredible advances have been made in the past decade with natural language processing, text to speech production, speech recognition, logical reasoning about the real world, image and sensor understanding, robotics, and navigation. Artificially intelligent assistants are becoming more common and more useful as they gain ability to communicate with each other, with computer-based systems and with the humans they serve.

It seems reasonable to anticipate that the human thirst for knowledge itself will be augmented increasingly by the use of pattern-recognizing software systems to the point that some aspects of discovery may be attributed to artificially intelligent programs that will sift vast amounts of data looking for correlations and anomalies adding to the sum of human knowledge. Whatever else the twenty-first century may bring, it will be a time of rapid innovation and discovery, rivaling centuries of the past and casting light into the future.

Finally, one cannot help but be concerned about the preservation of the vast quantities of digital information that will be produced in the decades ahead. Digital media have uncertain lifetimes and the interpretation of binary information often requires specific, executable software to render the meaning of any preserved bits. Software that runs now may not run on future machines and future operating systems, leaving us with vast quantities of "rotten bits." Forestalling a potential digital dark age will be another of the challenges that lie ahead as we fashion a future filled with new digital services upon which our society will depend.

VINTON G. CERF

Preface

This textbook is targeted to postgraduate or advanced undergraduate students. Gone are the days when voice was the only telecommunications service. In the current digital world, broadband and wireless networks enable multiple services, which are continuously evolving. Therefore, an updated textbook is needed that presents the rapidly evolving communications infrastructure to date and the exciting services that it supports. This is drastically different from the old style books on the telco infrastructure focusing on telco-centric network and services such as PSTN or B-ISDN.

This book presents the main products and services that are provided by current digital providers (Information and Communication Technologies players, including Internet behemoths, telecommunications operators, etc.). It covers services such as enriched communications, fixed and mobile broadband, financial services for unbanked in emerging markets, Pay TV, data communications for machines (also known as Internet of Things), digital home, and so on. This book is a complete and structured compendium of successful digital services with illustrative examples.

It should be stressed that this book's focus is on services paid for by customers (i.e., not those subsidized with advertising like social networks), but not the underlying technologies or internal capabilities to offer them that are not directly perceived by customers (therefore, out of scope are important topics of network management or a deep dive into future network technologies like Software Defined Networks and Network Function Virtualization, the future 5G networks, Fog computing architectures, etc.).

As opposed to technical-only textbooks, all topics in this book are addressed by taking into account the reason why customers demand the service, as well as the strategic and business perspective of how players can leverage competitive advantages to provide those services, that is, the underlying consumer and economic dynamics that determine the success or failure of service offerings.

By doing so, readers can get a better understanding of the key services in the highly dynamic Internet and telecom industries. They can get a holistic view of the main services from a strategic perspective. These up-to-date new areas have become extremely dynamic, and is still subject to continuous change.

The book is structured around six sections, each composed of one or more chapters, as follows:

I. Core telecommunications services
- Voice
- Broadband

- Convergence
- Devices

II. Telecom, cable and Pay TV becoming a single industry
- Pay TV

III. Services for enterprises
- IoT
- Cloud

IV. Emerging markets
- Unbanked

V. Other
- Value-added services
- MVNOs

VI. Innovation: This section shows other digital services that so far have not been adopted massively by paying customers; therefore, they are still labeled as innovation in this book.
- Digital home
- Videoconference and telework

Antonio Sánchez
Belén Carro

Acknowledgments

Our gratitude to all the people who have made this book possible. First of all, to the contributors who have helped with some of the chapters. Also, to the publisher (particularly Mary Hatcher and Divya Narayanan for their support) and the IEEE for giving us this opportunity. In addition, the reviewers who provided feedback to improve, at both the proposal stage and during the writing of this book.

List of Contributors

Jesus Llamazares Alberola

Jaime Bustillo

Joaquín M. Lopez Muñoz

Francisco Saez

Stefan Wesner

Chapter 1

The Evolving Voice Services: From Circuit Switching to Voice-Over LTE/FTTH)

Voice is the most pervasive telecommunication service, particularly mobile voice, with 4.7 billion unique mobile subscribers as of April 2016,[1] out of a total 7.8 billion mobile connections. Legacy circuit switched voice still accounts for majority of world telecom customers; however, voice-over IP is finally starting to replace it. Voice-over IP technologies such as H.323, Session Initiation Protocol (SIP), and voice codecs including wideband, carrier grade infrastructure known as IMS-IP Multimedia Subsystem[2] have been available for more than a decade, but deployment of All-IP access networks (FTTH (Fiber-to-the-Home) and LTE (Long-Term Evolution)) has been the trigger for their adoption. Only in the enterprise segment, the legacy PBX systems have initiated the migration earlier. In fixed networks, initial FTTH deployments maintained simultaneous use of legacy copper for voice. In 4G mobile networks, fallback to 2G/3G for voice has been a temporary patch until VoLTE deployment in the network and progressive availability of enabled devices. HD Voice is included in VoLTE, but is also available in 3G networks without VoIP. Voice-over WiFi is emerging as a complement mainly for zones with poor cellular coverage.

1.1 CUSTOMER NEED: REMOTE COMMUNICATION

Voice has been the dominant, or even only, telecommunications service for decades. The possibility of talking to people in remote places has been, and still is, a killer application. It has substituted more primitive means to communicate, both non-real-time and real-time. Furthermore, the convenience of talking wherever you are, thanks to the more recent mobile telephony that has certainly surpassed the

Digital Services in the 21st Century: A Strategic and Business Perspective, First Edition.
Antonio Sánchez and Belén Carro.
© 2017 by The Institute of Electrical and Electronics Engineers, Inc. Published 2017 by John Wiley & Sons, Inc.

success of fixed telephony, allows one to communicate in more remote areas without terrestrial infrastructure.

1.2   FTTH VOICE

The penetration of FTTH subscribers is still nascent; for example, in Europe there were almost 36 million (including Fiber to the Building) in Europe at the end of September 2015.[3] This has triggered the adoption of voice-over IP in fixed lines, although initial FTTH deployments supported simultaneous use of legacy copper for voice. Typically, the FTTH customer premises equipment, Optical Network Terminal (ONT), includes a voice port (RJ-11 connector) that interfaces with the legacy internal copper network or directly with the legacy telephony terminal. The usage is transparent for customers, since they make and receive calls normally with their legacy endpoints. The FTTH device includes a gateway that converts analog telephony into voice-over IP, which in turn is connected to the VoIP infrastructure of the operator.

1.3   VOICE-OVER LTE (VoLTE)

Voice-over LTE is the name given to the technology that provides telephony to 4G customers. It is based on voice-over IP, since LTE gets rid of circuit switched voice. In order to guarantee voice quality, VoLTE traffic is prioritized versus other kinds of traffic, which is also IP. Initial LTE deployments did not include VoLTE support, resorting to Circuit Switched FallBack (CSFB), which is based on disconnecting voice from the 4G network and connecting it to a legacy network (3G or even 2G) to establish the voice call. Once the call is finished, 4G connection can be set up again.

As of the end of the first quarter of 2016, there were 58 operator launches, in 32 countries,[4] with 228 end user devices. This was still a relatively small percentage of the total number of LTE networks: 467 operators in 153 countries, reaching 48% of the population. As could be expected, these countries are mainly in Europe, North America, South East Asia, and Oceania:

- Asia Pacific:
  - Japan by NTT Docomo, KDDI, and Softbank (the "big three")
  - South Korea by Korea Telekom (KT/KT Powertel), SK Telekom, and LG Uplus
  - Hong Kong by China Mobile, 3 (Hutchinson), CSL (HKT), and SmarTone
  - Singapore by SingTel, StarHub, and M1
  - Taiwan by Hon Hai, Asia Pacific Telecom, and Taiwan Mobile
  - Australia by Telstra and Vodafone
  - China by China Unicom and China Mobile (not China Telecom)
  - Others: Indonesia by XL (Axiata) and SmartFren; Thailand by AIS, True-Move, and Telenor (DTAC); Kuwait by KTC; Cambodia by Southeast Asia Telecom, and others

- Europe:
  - France by Bouygues and Orange
  - Germany by Deutsche Telekom, Telefonica,[5] and Vodafone (all three mobile network operators)
  - Italy by Telecom Italia and Vodafone
  - The United Kingdom by 3 (Hutchinson) and Everything Everywhere (BT)
  - Spain by Vodafone[6]
  - Others: Austria by Telekom Austria; Czech Republic by Deutsche Telekom; Denmark and Norway by Telenor; Switzerland and Liechtenstein by Swisscom; The Netherlands by Tele2; Portugal by Vodafone; Romania by Orange[7]; Russian Federation by Vimpelcom, and others
- North America
  - The United States by AT&T, Deutsche Telekom, Verizon Wireless, and Evolve Broadband (big four except Sprint-Softbank)
  - Canada by Rogers and Bell
- Africa: Tanzania, Uganda, and Nigeria by Smile; South Africa by Vodacom (Vodafone) . . .

It can be seen that all major mobile telecommunication groups in the world have launched VoLTE in at least one market. Among the exceptions are

- America Movil in Latin America
- Etisalat and Saudi Telecom in Middle East
- China Telecom in China
- Bharti Airtel in India
- Telekomunikasi Indonesia and Chungwa Telecom in Asia
- Telia in Europe
- MTN in Africa
- Telus in Canada
- . . .

However, it has to be noted that as of April 2016, there were already 126 operators investing in VoLTE deployments, trials, or studies in 60 countries.[8]

In terms of adoption, the operator that is most further along is probably AT&T, which launched in mid-2014, and as of the end of 2015, has more than 27 million VoLTE subscribers (the highest in the United States), and has a coverage for 295 million Americans.[9]

In November 2015, South Korea became the very first country to provide commercially interoperable VoLTE service among operators.[10] Before this milestone, customers could only enjoy the superior quality of VoLTE when calling other customers of the same operator. There are three operators (KT, SKT, and LG Uplus) with around 35 million VoLTE-enabled subscribers that are fully commercially interoperable. By that time, South Korea has marketed 90 different models of VoLTE-enabled handsets in the country. It has to be noted that the three operators were also the first in the world to launch their respective standalone services in December 2012, with the next country launching in May 2014 only.

Also a Korean operator achieved another major world milestone in terms of interoperability—in this case, an international roaming service. At the end of 2015, South Korea's KT and Japan's NTT DOCOMO offered a bilateral VoLTE roaming service. Previous milestones were unilateral (e.g., Korean LG Uplus with Japanese KDDI). Other mobile network operators involved in VoLTE roaming trials include China Mobile, KPN, SK Telecom, and Verizon Wireless.[11]

In terms of end user devices, the number includes carrier and frequency variants. As of June 2015 (219 VoLTE capable out of 3253 LTE devices), 198 smartphones had been announced by all the leading vendors, including Apple, Asus, Fujitsu, HTC, Huawei, LG, Motorola, Pantech, Samsung, Sharp, and Sony Mobile.[12]

1.4 VOICE-OVER WiFi

Voice-over WiFi (VoWiFi) is the term used by operators in order to refer to the provision of telephony services over a WiFi access. From the user perspective, the experience is similar to VoLTE, or traditional voice, that is, just call and answer. However, from the network perspective, it is quite different, since the WiFi network is not an integrated part of the mobile network. And therefore, the quality of the call can be poorer, and in general similar to that of over-the-top (OTT) applications.

Compared to VoLTE, the number of VoWiFi launches is much more limited, just 18 operators in 11 countries, by the end of March 2016. Also, deployments have been much more recent, starting only in September 2014 (in the United States by T-Mobile/Deutsche Telekom). In the United States, all four major operators have launched VoWiFi by now. Other countries with more than one operator include the United Kingdom, Switzerland, Hong Kong, and Canada. The remaining markets are China, Czech Republic, Denmark, Liechtenstein, South Africa, and Thailand. It has to be noted that most operators that have launched VoWiFi in a given market have also launched VoLTE in the same market. An exception is Sprint (Softbank) in the United States.

There have been operators that have launched VoWiFi service, without integrating it tightly with the VoLTE/IMS infrastructure and devices. One example is Telefonica in the United Kingdom and Latin America, with its service TU (formerly TU Go) that can be considered a pioneer in the world (perhaps together with Rogers in Canada), since it was launched in the United Kingdom in March 2013 (and afterward in Argentina, Peru, Brazil, and Colombia).[13] In the first quarter of 2016, it announced the integration with iPhone's native WiFi calling feature in Brazil.

Another announcement worth mentioning is the collaboration between Google and mobile network operators (19 major ones) in the field of Rich Communications Suite (RCS). Google will provide the user experience for Android devices (including an open-source version of the client), initially for messaging, but with support for advanced calling features, which among other things also include calling over WiFi, in the future.[14]

1.5 HIGH-DEFINITION (HD) VOICE

High-Definition voice provides superior voice quality based on wideband codecs, which (as opposed to legacy telephony) include more frequencies (bandwidth) of the analog sound (legacy narrowband reaches 3 kHz, whereas human voice reaches 14 kHz). HD voice has been available for quite long time, for both voice-over IP applications and circuit switched telephony (particularly mobile, even before 4G, with 3G and even 2G[15]). Therefore, the number of commercial services is more widespread, with 162 operator launches in 89 countries as of the end of the first quarter of 2016, including VoLTE deployments.

1.6 OVER-THE-TOP SUBSTITUTES

The term "over-the-top" refers to services that are offered independent of the underlying network. One of the most prominent consumer over-the-top applications for PC has been Skype, which is now part of Microsoft. In January 2016, the company reported that Skype has over 900 million downloads on iOS and Android.[16] It also has an enterprise counterpart, called Skype for Business, with a major update released in December 2015, including voice telephony without PBX, and online meetings for up to 10,000 participants.

Perhaps the most prevalent over-the-top application for mobile is WhatsApp,[17] which was started in 2009. In early 2014, WhatsApp was acquired by Facebook for $16 billion ($4 billion in cash, and around $12 billion in shares).[18] In February 2016, it reached the astonishing milestone of 1 billion monthly active users, although it has to be noted that its predominant usage is for messaging. It is worth mentioning that WhatsApp was an inexpensive app, but in early 2016 the company announced that it was becoming free, that is, no longer charging any subscription fees; its usage is also free in free WiFi networks, but of course it consumes data allowance if used with mobile broadband. The company also confirmed that it will not include advertising to subsidize its costs, but rather test business models around commercial messages (communications with businesses). In April 2016, it included end-to-end encryption for all the information exchanges, including voice. The app is available for iOS, Android, and Windows Phone. By the end of 2016, it will end its support to minority or legacy operating systems like BlackBerry, Nokia S40/Symbian S60, as well as older versions of Android 2.1 and 2.2 and Windows Phone7.1. In the first quarter of 2015, WhatsApp started rolling out voice-over IP calls, progressively in different operating systems; this feature was already available in Facebook Messenger application, which by then accounted for more than 10% of mobile VoIP calls globally according to the company.[19] By mid-2016, 100 million voice calls per day were made through WhatsApp.[20]

ACRONYMS

2G	2nd Generation
3G	3rd Generation
4G	4th Generation
CSFB	Circuit Switched FallBack
FTTH	Fiber-to-the-Home
HD	High Definition
IMS	IP Multimedia Subsystem
IP	Internet Protocol
LTE	Long-Term Evolution
ONT	Optical Network Terminal
OS	Operating System
OTT	over-the-top
PBX	Private Branch Exchange
RCS	Rich Communications Suite
RJ	Registered Jack
SIP	Session Initiation Protocol
VoIP	voice-over IP
VoLTE	voice-over LTE
VoWiFi	voice-over WiFi
WiFi	Wireless Fidelity

NOTES

1. GLOBAL DATA: Mobile connections, including M2M/Unique mobile subscribers, Apr 2016. GSMA Intelligence. https://gsmaintelligence.com/ (accessed April 10, 2016).
2. A. Sánchez-Esguevillas, B. Carro, G. Camarillo, Y. B. Lin, M. A. García-Martín, L. Hanzo (2013) IMS: the new generation of Internet-protocol-based multimedia services. *Proceedings of the IEEE,* IEEE. http://ieeexplore.ieee.org/xpl/login.jsp?tp=&arnumber=6479674 (March 14, 2013).
3. Croatia, Germany, and Poland joined the FTTH ranking. Fibre to the Home Council Europe (iDate). www.ftthconference.eu/images/Banners/Conference2016/News/PR20160217_FTTHranking_panorama_award.pdf (February 17, 2016).
4. Delivering an all-IP world. GSMA, July 26, 2016. www.gsma.com/network2020/resources/all-ip-statistics/ http://www.gsma.com/network2020/wp-content/uploads/2016/04/Network-2020-VoLTE-launches-31-March-2016.pdf (updated June 30, 2016).
5. Faster and better telephony with voice-over LTE (VoLTE) within the entire 02 network.O2-Telefonica. www.telefonica.de/fixed/news/5844/faster-and-better-telephony-with-voice-over-lte-vo-lte-within-the-entire-02-network.html (April 16, 2015).
6. Vodafone España ofrece voz 4G -VoLTE- en toda su red 4G+. Vodafone. www.vodafone.es/conocenos/es/vodafone-espana/sala-de-prensa/notas-de-prensa/vodafone-espana-ofrece-voz-4g–volte–en-toda-su/ (July 7, 2015).

7. Orange Romania launches VoLTE; Wi-Fi calling to follow. www.telegeography.com/products/commsupdate/articles/2015/09/14/orange-romania-launches-volte-wi-fi-calling-to-follow/ (September 14, 2015).

8. VoLTE global status. GSA. http://gsacom.com/paper/volte-global-status/ (April 8, 2016).

9. AT&T's voice over LTE network reaches more than 27 million subscribers. AT&T. http://about.att.com/innovationblog/122915voiceoverlte (December 29, 2015).

10. GSMA welcomes launch of world's first commercial interconnected VoLTE service in South Korea. GSMA/South Korea's interconnected VOLTE service lifts off. www.gsma.com/newsroom/press-release/gsma-welcomes-launch-of-worlds-first-commercial-interconnected-volte-service-in-south-korea/ http://www.gsma.com/newsroom/all-documents/south-koreas-interconnected-volte-service-lifts-off/ (June 18–December 1, 2015).

11. KT Corporation and NTT Docomo land world's first with launch of bilateral VoLTE service. GSMA. www.gsma.com/newsroom/blog/kt-corporation-and-ntt-docomo-land-worlds-first-with-launch-of-bilateral-volte-service/ (October 20, 2015).

12. GSA confirms 3253 LTE devices, LTE-Advanced takes hold. GSA. http://gsacom.com/press-release/gsa-confirms-3253-lte-devices-lte-advanced-takes-hold/ (June 21, 2015).

13. Now your mobile number works on Wi-Fi across your devices. so you're always connected / Happy 3rd Birthday TU Go UK!/Wi-Fi Calling with TU Go in Brazil. Telefonica. https://tu.com/en/https://tu.com/en/weblog/2016/03/07/happy-3rd-birthday-tu-go-uk/ / https://tu.com/en/weblog/2016/03/30/wi-fi-calling-tu-go-brazil/ (accessed April 30, 2016, March 30, 2016, March 7, 2016).

14. Global operators, Google and the GSMA align behind adoption of rich communication services. GSMA. www.gsma.com/network2020/digest/global-operators-google-and-the-gsma-align-behind-adoption-of-rich-communication-services/http://www.gsma.com/newsroom/press-release/global-operators-google-and-the-gsma-align-behind-adoption-of-rcs/ (February 21, 2016).

15. 150 mobile operators launched HD voice service in 87 countries. GSA. http://gsacom.com/paper/150-mobile-operators-launched-hd-voice-service-in-87-countries/ (December 15, 2015).

16. Earnings Release FY16 Q2. Microsoft. www.microsoft.com/en-us/Investor/earnings/FY-2016-Q2/press-release-webcast (January 28, 2016).

17. WhatsApp blog. WhatsApp. https://blog.whatsapp.com/?l=en&set=yes (April 5, 2016).

18. Facebook to acquire WhatsApp. Facebook.http://newsroom.fb.com/news/2014/02/facebook-to-acquire-whatsapp/ (February 19, 2014).

19. Facebook Q1 2015 earnings call transcript. Facebook. http://files.shareholder.com/downloads/AMDA-NJ5DZ/1862880422x0x823326/A88E8ECF-8532-4F35-B251-4F95C2A4C6B3/FB_Q12015_Transcript.pdf (April 22, 2015).

20. WhatsApp calling: 100 million conversations every day. WhatsApp. https://blog.whatsapp.com/10000625/WhatsApp-Calling-100-million-conversations-every-day (June 23, 2016).

Chapter 2

Internet Services: From Broadband to Ultrabroadband

From the old times of very low-speed narrowband fixed dial-up and mobile 2G GPRS, telecom operators have been offering broadband access massively. The mobile network, being a shared medium, still lags the fixedline network in speed today, but in both cases the available speeds are already quite high and continue to increase.

The mobile network surpasses the fixed line network in terms of adoption. In 2015 there were already more than 3 billion mobile Internet subscribers (unique).[1] This figure represents more than half of mobile subscribers (which includes those without Internet: almost 5 billion unique mobile subscribers—and 8 billion total mobile connections—by the end of 2016[2]). Related to access device, smartphones already represented 45% of total mobile connections by the end of 2015.[3] In what respects to technology, LTE (4G) achieved 1 billion connections milestone in 2015[4] (www.gsmaintelligence.com/, last accessed January 12, 2017).

In fixed broadband xDSL, copper-based technologies offer speeds that depend on the distance to the central office. Typical speeds are asymmetrical with downlinks having a few Mbps (megabits per second), increasing to several tens of Mbps with VDSL for closest buildings (or with fiber to an intermediate point—cabinet—in order to reduce the distance). Uplink is typically up to 1 Mbps. On the other hand, cable operators have been upgrading their networks by deploying fiber very close to customer premises, the so-called Fiber-to-the-Node, with coaxial cable left only to the last few meters to tens of meters. Based on DOCSIS (Data Over Cable Service Interface Specification) 3.0 specification, the cable network provides commercial speeds of several hundred Mbps and could reach up to 10 Gbps (gigabits per second) in the downstream and 1–2 Gbps in the upstream with improved modulation of DOCSIS version 3.1. or even 10 Gpbs symmetrical with full-duplex techniques, which is still in the R&D stage.[5] Finally, FTTH is an All-IP network, which as its name implies deploying fiber end to end (central office to customer premises, home, or building plus LAN (Local Area Network)). Commercial speeds have

Digital Services in the 21st Century: A Strategic and Business Perspective, First Edition.
Antonio Sánchez and Belén Carro.

already reached 10 Gbps symmetrical since 2014,[6] although with limited coverage, and 1–2 Gbps symmetrical with broader coverage. In any case, technology could in principle provide higher speeds. In terms of coverage, there were (as of September 2015) 53 countries (with at least 200,000 households) that have at least 1% of homes subscribing to fiber (FTTH or Fiber-to-the-Building plus LAN), with top 6 in Asia (all above 50% household penetration).[7]

As opposed to fixed lines, mobile speed is more limited. Mobile networks are migrating from 3G to 4G, with main improvements in speed/capacity but also reduction in latency. 4G deployments are quite widespread, although the population coverage is not universal yet. The latest generation is LTE-Advanced (available in around 50 countries[4]) with the so-called carrier aggregation, which combines larger spectrum to increase speed. The maximum commercial advertised speed is 600 Mbps in downlink in Australia with the so-called Category 11 devices (under 3rd Generation Partnership Project (3GPP) standard).[8] The modem supports 150 Mbps uplink, with new modems supporting even 1 Gbps downlink[8] announced in early 2016 and commercial in second half 2016. However, it should be noted that the real speeds are much lower, given the fact that capacity is shared among all users connecting to the same base station.

Finally, operators are also offering Internet connectivity through WiFi, both in the fixed access (routers with WiFi) and in mobile access (offering multiple hot-spots, sometimes through partnerships). With the new large available spectrum in 5 GHz and the recent 802.11ac standard, both capacity and speed have increased significantly.

Since broadband has a very positive impact on society, many governments are fostering universal coverage (e.g., Europe's ambition of providing 30 Mbps universally by 2020[9]).

2.1 CUSTOMER NEED: CONNECTIVITY AND SOCIAL INCLUSION

Since the inception of the Internet, connectivity has been embraced by billions of people, and even more to come, with availability and affordability the only limiting factors. Initially, people started to connect through dedicated systems (e.g., in government organizations) and then through modem dial-up, available to those with a fixed phone line. Given the fact that mobile lines are now much more widespread, the availability of mobile Internet made the figures skyrocketed.

Connectivity is the basic building block for any kind of digital services, some of which may not have been invented yet. Social networks and messaging are just two examples of applications based on connectivity that have surpassed the figure of 1 billion active users monthly.

Beyond this, connectivity can now be considered a social right, and that is why overcoming the digital divide is a top priority for governments. Connectivity is a proven mean of social inclusion, similar to enabling access to other basic rights like health and education. Along this line, there are rigorous studies that show the positive impact of connectivity in general and more specifically broadband in the Gross Domestic Product of an economy.[10]

2.2 FIXED LINES: DEPLOYING FIBER CLOSER TO CUSTOMER PREMISES: xDSL, CABLE, FTTH

Access networks, as well as mobile networks, have rapidly evolved toward broadband. It has come a long way since the xDSL[11–13] family started as a provisional upgrade of PSTN (Public Switched Telephone Network) and ISDN[14] (Integrated Services Digital Network) narrowband access, and xDSL technology still survives but is being quickly replaced by other wired and wireless technologies.

Generally, at home, broadband access networks are wired: The xDSL copper wire family, the Hybrid Fiber-Coaxial (HFC) network, with fiber cores and coaxial distribution with a trend to evolving toward a complete fiber network,[15] and the FTTH[16] (Fiber-to-the-Home) networks. Also, mobile networks might be considered among broadband access networks because of their increasingly higher transmission rate, their wide coverage and a minimal amount of network installation, and the wide use of mobile terminals like smartphones.

A wired network is composed of a series of cables and other equipment between the local exchange of the operator and the network termination point that separates the client premises from the access network. The access network is the most important and expensive property of the operator, constantly evolving due to new service offers that imply new requirements such as more bandwidth, less latency, and so on, as well as regulatory changes.

2.2.1 xDSL

xDSL (x Digital Subscriber Line) includes a family of Internet broadband access technologies based on the telephone subscriber loop (copper wire) digitalization. The main advantage of the xDSL family has traditionally been the reutilization of already deployed infrastructure with almost 100% coverage, partially or totally amortized. xDSL access is based on the conversion of the PSTN copper line to a high-speed digital line capable of supporting broadband services simultaneously with voice (with a channel located between 300 Hz and 3–4 kHz) with a minimum amount of interference. It should be noted that ISDN is not compatible with xDSL since it employs the xDSL lower frequency band and their spectra would be overlapped.

2.2.1.1. ADSL

ADSL (Asymmetric Digital Subscriber Line) is the most common xDSL access technology, in which the uplink (from user to local exchange) and downlink (from local exchange to user) transmission rates are different due to the classic need for higher speed in the downlink demanded for traditional services like web browsing, video streaming, or email.

The main components of an ADSL link are the following:

- The ADSL Terminal Unit-Remote (ATU-R) or modem in the user premises
- The ADSL Terminal Unit-Central (ATU-C) or modem located in the telco premises, grouped in a DSLAM (Digital Subscriber Line Access Multiplexer) in the local exchange

A splitter must be placed before each modem. The splitter is a set of two filters: a high-pass filter and a low-pass filter that separate the low-frequency (telephony) and high-frequency (ADSL) signals.

ADSL employs Discrete MultiTone (DMT) modulation that is similar to OFDM (Orthogonal frequency-division multiplexing). It uses multiple subcarriers, each QAM modulated, with a separation of 4.3125 kHz and a bandwidth of 4 kHz for each subcarrier. When establishing the connection between the ATU-R and the ATU-C, the SNR is measured for each subcarrier band, and then the data flow is distributed depending on the SNR value: When the subcarrier SNR is high, it will transmit a higher data rate. DMT is a complex modulation scheme that enables the success of copper wire to transmit high data rates from the initial filter-limited 4 kHz used in traditional voice, but the modulation algorithm is translated into an IFFT (inverse fast Fourier transform) in the modulator and an FFT (fast Fourier transform) in the demodulator located on the other extreme of the loop making the DMT simpler to execute. Besides, these operations may be easily carried out developing the modem core on a DSP (Digital Signal Processor). DMT has two versions: DMT with FDM (frequency-division multiplexing) and DMT with echo cancellation. In DMT with FDM, the spectra of uplink and downlink signals do not overlap, and the modems are simpler, but the downlink transmission rate is reduced since the lower frequency subcarriers are not available. Lower frequency subcarriers are more desirable because the copper wire attenuation is lower there. In DMT with echo cancellation, an echo canceller separates the signals from both transmission directions, so more downlink transmission speeds are achieved in exchange for a more complex modem design.

The length of the local loop limits the maximum achieved ADSL binary rate since the longer the loop length, the higher the overall attenuation for the transmitted signals. Also, a higher frequency implies more attenuation per unit length. This is why the distance from the user to the local exchange should not exceed 2.5 km to guarantee the quality of service. It should preferably be under 1.5 km, which is the value from which the link data rate begins to drop.

ADSL has been upgraded several times (e.g., ADSL2 and ADSL2+) to increase its uplink and downlink speeds, making it very competitive compared to other wired access solutions like HFC and FTTx. This has also enabled the operators to offer integrated packs, including voice, data, and television, known as Triple Play. The new ADSL versions incorporate advanced modulation schemes and physical resources management that increase the basic ADSL capacity up to 24 Mbps in the downstream and include improvements to avoid noise or interference and reduce the effects of attenuation reaching distances of up to 9 km.

2.2.1.2. VDSL

VDSL stands for Very-High-Bit-Rate Digital Subscriber Line and may be seen as a natural evolution of ADSL, offering symmetric (26 Mbps) or asymmetric speeds (52 Mbps for downlink and 16 Mbps for uplink). It increases the frequency bands used to transmit the data: four channels versus the two channels employed in ADSL—two for the uplink and two for the downlink, accompanied by an increment of the transmitted power.

VDSL can be used for transmitting HD television by compressing the video and incorporating FEC (Forward Error Correction) mechanisms to achieve low error rates. It usually employs DMT modulation but QAM/CAP (Quadrature Amplitude Modulation/Carrierless Amplitude/Phase) modulation is also used, plus TDM/TDMA scheme.

2.2.1.3. VDSL2

VDSL2 stands for Very-High-Bit-Rate Digital Subscriber Line 2, as an evolution of VDSL, designed to support the Triple Play services, including voice, video, data, HDTV (High Definition TV), and interactive gaming. It may transmit data symmetrically or asymmetrically, and the transmission speed is highly dependent on the distance to the local exchange: 250 Mbps at the local exchange output decreases to 100 Mbps at 1 km distance and 50 Mbps at 2 km. The decrease becomes slower at longer distances. For short local loops, symmetry is easily achieved reaching over 100 Mbps under suitable conditions. In fact, high speeds are only achieved for distances of up to 400 m from the local exchange, so these connections are normally served by a fiber optic node located on the street. Therefore, combining copper wire and fiber optic the VDSL/VDSL2 coverage can be widened, otherwise only subscribers close to the local exchange could enjoy high speeds.

The main difference between ADSL and VDSL is the available bandwidth: ADSL and ADSL2 use a 1.104 kHz band divided into 256 carriers, ADSL2+ employs a 2.208 kHz band divided into 512 carriers, and VDSL may use bands of 8, 12, and 17 MHz or, in case of VDSL2, 30 MHz, thus enabling higher transmission speeds.

2.2.2 FTTH

FTTx (Fiber-to-the-x) denotes the group of technologies that employ fiber optic (more or less) close to the subscriber's premises:

- *FTTH (Fiber-to-the-Home):* The fiber reaches the inside or the façade of the client's house or office. It may be point-to-point, with one or two fibers from the central office to the user facilities, or point-to-multipoint, with one fiber from the central office shared by multiple users.

- *FTTB (Fiber-to-the-Building):* The fiber optic typically ends in an intermediate distribution point inside or around the subscribers' building. From this intermediate distribution point, the users receive the service through VDSL2 over copper wire or Gigabit Ethernet over CAT5 twisted pair, enabling the fiber deployment to be progressive, saving time and money and reutilizing the subscriber's infrastructure.

- *FTTC (Fiber-to-the-Curb):* Reaching the house with xDSL.
- *FTTN (Fiber-to-the-Node or Neighborhood):* The fiber ends farther away from the user than FTTH and FTTB, typically near the neighborhood.

Several technological solutions exist for offering FTTx:

- PON (Passive Optical Networks) do not require active electronic components between the user and the central office.
- ASON (Automatically Switched Optical Network) require active electronic components between the user and the central office.

PON and specially GPON (Gigabit PON) technologies are the most widely deployed technologies: Since no active electronic or optoelectronic devices are needed to connect the central office to the user, its investment and maintenance costs are considerably lower than ASON technologies.

2.2.2.1. PON

PONs compete and complement xDSL, HFC, and fixed wireless networks such as WiMAX. It consists only of passive components: fiber links, splitters, and couplers. It is a point-to-multipoint topology.

Several types of PONs exist; they share the general technology but differ in the specifications and in the higher layer protocols.

- *APON (ATM based PON):* It uses ATM (Asynchronous Transfer Mode) encapsulation for data transport.
- *BPON (Broadband PON):* APON successor, it also employs ATM encapsulation and presents enhanced features like higher speed. It is an ITU-T (International Telecommunication Union-Telecommunication Standardization Sector) standard.
- *EPON or GE-PON (Ethernet PON or Gigabit Ethernet PON):* It uses Ethernet for data encapsulation. IEEE (Institute of Electrical and Electronics Engineers) standard.
- *Mid-GPON or Giga PON:* It uses a new encapsulation system (Generic Encapsulation Method (GEM)) that supports ATM, Ethernet, and TDM for data transport.

2.2.2.2. GPON

GPON is a set of ITU-T recommendations collected in G.984.x where techniques to share a common medium (the fiber optic) among users, encapsulate information, and manage the network elements are described.

A GPON network consists of the following:

- *OLT (Optical Line Terminal):* Located in the telco's facilities, it consists of several GPON line ports, each one supporting up to 128 ONT (typically 64). Some systems may allocate over 7.000 ONTs in the same space occupied by a DSLAM in xDSL.
- *ONTs (Optical Networking Terminals):* Located in the FTTH subscriber facilities. In FTTN, ONTs are substituted by MDUs (Multi-Dwelling Units) that offer VDSL2 to reach the users' house, accomplishing the short distance needed to get the symmetrical 100 Mbps data rates for each user.

GPON shows a point–multipoint or tree topology where splitters play an important role.

- *In the downlink:* A splitter divides the light signal in its input among several outputs, enabling the downlink traffic from the OLT to be distributed among the different users. It may be several passive splitters in the form 1xn, where n = 2, 4, 8, 16, 32, 64, or 128 in different places until the clients are reached. In case of using GPON extenders with a PON regenerator or an active optic amplifier between the OLT and the splitter, distances of up to 60 km may be reached with a division factor of 128, which is adequate for rural areas. In the downlink, an optic broadcast is performed but each ONT will only process the traffic that corresponds to it, thanks to the AES security techniques.
- *In the uplink:* The traffic from the ONT to the OLT is aggregated by the same splitter that acts as a combiner in the opposite traffic direction. Upstream and downstream traffic share the same fiber optic but is distributed on a different wavelength to avoid collisions between the two traffic directions. The wavelengths are typically 1490 nm in the downlink and 1310 nm in the uplink. In case of broadcasting video, the 1550 nm wavelength may be used through WDM (*wavelength-division multiplexing*). TDMA (time-division multiple access) guarantees a transmission free of collisions from the ONT or MDU to the OLT, and the transmission only takes place when needed to avoid transmission inefficiency.

Video may be offered in two different ways simultaneously:

- *RF (radiofrequency):* Cable operators may gradually migrate to IPTV (IP television). ONTs have a RF coaxial output connected to the STB (set-top box).
- *IPTV:* The video signal is transformed by the central headend into an IP data flow, transmitted over the same IP link with higher data priority than Internet traffic. The STB, connected with Gigabit Ethernet to the ONT to the remote gateway (residential gateway), will again convert the IP data flow back into a video signal.

GPON equipment incorporates advanced QoS (Quality of Service) and IP multicast capacities, allowing the operators to offer several HD IPTV channels and interactive and customized services.

The GPON operation is simple. From the OLT, a fiber enters a splitter dividing the signal into fibers and then end on up to 64 ONTs. The splitters may be one-stage (1×64) or cascaded, for example, 1×4 entering into four 1×16. The GPON uses a method called ranging in which the OLT coordinates the upstream transmission of every ONT in each timeslot. The delay in each ONT must be lower than the guard time to avoid collisions. If an ONT does not work properly, it will affect the whole downlink. Also, the aging of optical components may require constant monitoring precisely by the OLT. Transmission speeds, frame format, optical interface specifications, and so on depend on the type of PON.

GPON offers remarkable advantages over its predecessor PON technologies:

- *Higher speed:* APON (ATM PON) and BPON (Broadband PON) offered lower speeds and were based on ATM, which added cost and complexity.
- *Voice, data, and video convergence over the same IP infrastructure:* With advantages for the operators in terms of, simplicity, flexibility, higher revenue, and the ability to accommodate current and future services to increase client loyalty, plus less CAPEX (CAPital EXpenditure) and OPEX (OPerational Expenditure).
- *GEM (GPON Encapsulation Method):* It supports any type of service (Ethernet, TDM, ATM, etc.) offering a higher bandwidth and being much more efficient. Also, operators may continue offering their traditional services (ATM-based voice, leased lines, etc.) without changing the equipment installed in user premises.
- GPON implements advanced OAM (Operation Administration and Maintenance) capability, offering a powerful end-to-end service management, including error rate, alarms, events monitoring, and so on.
- The Dynamic Bandwidth Allocation (DBA) allows the operators to offer the users more traffic when needed when the network capacity is not fully occupied (no other users are employing their whole available bandwidth).

Currently, the most common data rate by the GPON equipment suppliers is 2.5 Gbps downstream and 1.25 Gbps upstream, although higher speeds of up to 10 Gbps are also being offered.

Regarding transmission, the signal that travels through the fiber optic suffers from attenuation that increases with the distance, the number of splices and connectors, and the splitters' attenuation. The key GPON optical parameters[17] are listed as follows:

- ONT and OLT mean launched power (minimum), typically 0.5 and 1.5 dBm, respectively.
- ONT and OLT mean launched power (maximum), typically 5 dBm each.
- ONT and OLT minimum receiver sensitivity, typically −27 dBm (at 2488 Mbps) and −28 dBm (at 1244 Mbps), respectively.
- ONT and OLT minimum overload, typically −8 dBm each.

- Downstream optical penalty, typically 0.5 dBm each.
- FEC attenuation, typically 2 dB.
- Guard margin, typically 2 dB.
- *Optical division (splitters):* It causes a proportional signal attenuation: −3 dB for each division, for example, −3 dB for a 1:2 splitter, −9 dB for a 1:8 splitter.
- *Typical attenuation for physical elements:* Fiber optic at 1310 nm (per km): −0.4 dB, fiber optic at 1550 nm (per km): −0.3 dB, fusion splice: −0.1 to −0.2 dB, mechanical splice: −0.5 dB, insertion loss (connectors): −0.3 dB to −0.5 dB.

For calculating the maximum possible distance between an ONT and the OLT, the following parameters must be considered:

- Minimum available power, sensitivity, FEC, and guard.
- The existing splitters' attenuation.
- The splice attenuation and the number of splices.
- The connector attenuation and the number of connectors.
- The fiber optic attenuation per kilometer.

Signal reflections may also occur reducing the power balance, due to critical points like dirty or damaged connectors and low-quality or badly installed mechanical connectors. Reflection positions and value must be identified for any repair.

2.2.2.3. NG-PON

A successful technology must be able to adapt to future requirements and be compatible with their predecessor technologies, for example, ADSL has evolved to improved technologies like ADSL2, ADSL2+, or VDSL being backward compatible. GPON evolves to NG-PON (Next-Generation PON) resulting in an increase in GPON bandwidth and coverage, maximizing the utilization of the installed passive optic network from the central office to the subscribers or ODN (Optical Distribution Network) since its cost is about the 75–85% of the total cost of offering broadband fiber optic to the users. NG-PON has two variants: NG-PON1 or XG-PON and NG-PON2 or WDM-PON:

- *NG-PON1 or XG-PON:* Employs TDM, reaches up to 10 Gbps, its optical distribution network (ODN) is GPON compatible.
  - *XG-PON1:* Same principle as GPON (downlink optical broadcast, uplink TDMA, etc.), with some differences in wavelength, better security, and energy saving. It allows GPON-XG-PON1 coexistence, supports 10 Gbps downstream and 2.5 Gbps upstream. It is initially the preferred technology for ONTs.
  - *XG-PON2:* Supports 10 Gbps in both directions.

- *NG-PON2 or WDM-PON:* Simple technology that preserves the same TDM-PON point to multipoint architecture at the physical layer. Each ONU (Optical Network Unit)/ONT has a dedicated wavelength at the physical level. From a logical point of view, each wavelength may be seen as a point-to-point channel that will transport dedicated and symmetrical speeds to every user, from 100 Mbps to 10 Gbps. For transmitting over a single set of fiber optics without interference, different wavelength bands are used in the uplink and in the downlink. The uplink and downlink wavelengths may be unique for the FTTH residential or business client over one ONT, or else be shared by several FTTB/C clients through a MDU. In WDM-PON, the ODN of TDM-PON does not remain intact: the GPON and XG-PON splitter is substituted by an AWG (Arrayed Wavelength Grating), a passive component. While the GPON splitter replicates the optical signal through all its outputs from the central to the users, thus dividing power among all the outputs, the AWG directs each wavelength to its corresponding ONU with very low loss. For example, a 1:64 splitter implies a loss of power of 20 dB versus 8 dB introduced by an AWG. The extra optical budget may be used to reduce the optical components cost, or else to increase the split ratio or the distance covered. WDM-PON is able to support distances of up to 85 km without extenders, enabling operators to consolidate the active equipment in the access network and to noticeably reduce the number of local exchanges.

WDM-PON offers several advantages over GPON:

- The wavelength is not shared over time: Offering highly guaranteed different bandwidth, symmetric or asymmetric, dedicated and with no containment, to every user is simpler with WDM-PON.
- High bandwidth scalability, due to the bit rate transparency and the easy addition or elimination of channels.
- Greater distances and division factors due to the lower optical losses.
- Easier network operation, management, and maintenance.
- Higher security due to the traffic separation among clients.
- Easier to create open optical networks with wavelengths unbundling, allowing several operators to share the same physical access network, in the same way as current xDSL copper networks.
- Lower latency, important for applications like LTE (Long-Term Evolution) mobile hackhaul. It will also improve the user experience on online gaming, cloud computing services, unified communications, and so on.

2.2.2.4. TWDM-PON

TWDM (Time wavelength-division multiplexing)-PON is chosen as the primary solution for Next-Generation Passive Optical Network stage-2 (NG-PON2) architecture, thanks to its evolving specifications, wavelength plans, loss budgets, and key technologies that enable tunable ONUs. TWDM-PON allows different wavelengths to be assigned to different services. The following are its advantages:

- It offers a higher bandwidth of up to 10 Gbps to each user.
- Optimal flexibility in the bandwidth per user, fiber optic management, services convergence, and resources sharing.
- GPON compatibility: It may coexist and be developed over the current GPON deployment.
- No impact on external plant: It uses the same optical splitters to simplify the fiber optic management, ease the adjustments, and maximize compatibility with network components and end user equipment.
- Simple introduction: It may be gradually introduced in the current FTTx deployments using integrated solutions for multiple wavelengths, or else starting with a specific wavelength and adding more wavelengths as the bandwidth demand increases. It may be extended on a fiber-to-fiber basis just by adding and configuring new cards in the central office.
- It adds flexibility, supporting the superposition of multiple services, groups of users, or organizations over the same fiber. For example, an operator may simplify its operations using dedicated wavelengths to isolate applications provided by its different business areas. Also, with TWDM, different wavelengths may be assigned to different operators.

2.3 MOBILE: 4G LTE/LTE-ADVANCED

2.3.1 Mobile Evolution

The 3rd Generation Partnership Project (3GPP)[18] is an international organization that groups several telecommunications standard development organizations to establish the standards for generations of commercial cellular/mobile systems, providing complete system specifications through its releases. For example, third-generation technology is covered in Releases 99, 4, 5, 6, and 7.

There has been a long evolution path[19–21] from first-generation (1G) mobile technology to current fourth-generation (4G[22]) and the under development fifth generation (5G).

1G (e.g., AMPS,[23] NMT, TACS) first established seamless mobile connectivity through introducing mobile analog voice services, with no data service available.

2G (e.g., GSM/GPRS,[24] D-AMPS, CDMAone) went digital and provided voice and simple data like SMS (Short Messaging System) or very low-speed data connections to the mass market, popularizing the cellular mobile communications worldwide. GSM was developed to carry real-time services, including data services, over a circuit switched modem connection. GPRS evolution introduced an IP-based packet switched architecture using the same air interface and access method, TDMA (time-division multiple access). In the last EDGE (Enhanced Data rates for Global Evolution) release data rates reach up to 1.2 Mbps.

3G (e.g., CDMA2000/EV-DO, WCDMA[25]/HSPA+,[26] TD-SCDMA) enables mobile broadband services, capable of providing over 60 Mbps data rates. UMTS[27] (Universal Mobile Terrestrial System) reaches higher data rates with a new access technology WCDMA (wideband code-division multiple access) that emulates a circuit switched connection for real-time services and a packet switched connection for data services. Continuous evolution of 3G brings it closer to 4G capabilities.

4G (e.g., LTE, LTE-A) enhances 3G capacity and data rates: The highest theoretical peak data rate on the transport channel is 75 Mbps for the uplink, and 300 Mbps for the downlink, using spatial multiplexing (4×4 MIMO and 20 MHz bandwidth). It is sometimes even higher than that for broadband fixed access solutions.

The key to the current success of 4G is the multimode 3G/LTE ability. 3G delivers ubiquitous voice services and global roaming and enables broadband access outside 4G LTE coverage, while 4G provides more data capacity for richer content and more connections.

2.3.2 LTE

LTE specifications can be found in 3GPP Releases 8 and 9. LTE technical features include the following:

- *Wider channels:* Flexible support for channels up to 20 MHz (including 100 resource blocks, each resource block of 180 kHz) enabled with orthogonal frequency-division multiple access (OFDMA).
- *More antennas:* Advanced multiple input multiple output (MIMO) techniques to create spatially separated paths, spatial multiplexing in the downlink (up to 4×4 antennas).
  - MIMO[28] is used to increase the overall bitrate by transmitting several different data streams on several different antennas, using the same resources in both frequency and time, separated only by using different reference signals, and to be received by several antennas. MIMO may be used when the signal-to-noise ratio (SNR) is high, that is, there is a high-quality radio channel. For situations with low SNR, the use of other types of multiantenna techniques to improve the SNR, for example, transmission diversity, is preferred
- Higher order modulation, up to 64 QAM.
- Simplified core network: An all IP network with flattened architecture resulting in less equipment per transmission.
- Low latencies: Optimized response times (lower than 5 ms) for both user and control plane improves user experience.

LTE is based on OFDMA[29] (orthogonal frequency-division multiple access) that subdivides the available bandwidth into a multitude of mutual orthogonal narrowband subcarriers that can be shared among multiple users. OFDMA leads to high peak-to-average power ratio (PAPR) requiring expensive power amplifiers with high requirements on linearity, increasing the power consumption for the sender that would lead to expensive handsets. Therefore, a different solution was selected for the uplink: SC-FDMA[30,31] (single carrier-frequency-division multiple access) that generates a signal with single carrier characteristics, hence with a low PAPR. Besides, two modes operate for 4G: FDD (frequency-division duplexing) and TDD (time-division duplexing). LTE standard enables common FDD/TDD products in such a way that TDD shares most of FDD design and standard. There is an inherent seamless FDD/TDD interoperability, and an even tighter FDD/TDD interworking is planned.

The LTE access network is called eUTRAN (enhanced UMTS Terrestrial Radio Access Network), and is simplified to a network of base stations, the evolved NodeB (eNB), generating a flat architecture. The intelligence is distributed among the base stations to speed up the connection setup and reduce the time required for a handover. In certain services such as online gaming, the connection setup time for a real-time data session is vital for the end user. The time for a handover is essential for real-time services where end users tend to end calls if the handover takes too long. Handover is established with UTRAN and with other non-3GPP technologies like CDMA 2000, WiFi, or WiMAX.

LTE is an all-IP network, therefore voice is carried over IP (VoIP), known as VoLTE[32,33] (voice-over LTE). However, LTE must allow 3G handover as 4G coverage is not 100%. Several solutions exist to address this issue:

- *CSFB (Circuit Switched Fall Back):* Voice calls make a 2G/3G handover from 4G.

- *SVLTE (Simultaneous Voice and Data LTE) or (?) Dual Standby:* Dual radio using one chip for voice and one chip for data. It is hardly employed.

- *SRVCC (Single Radio Voice Call Continuity)*: Voice on LTE with CS backup. Voice over LTE in areas of LTE coverage, handover to 2G/3G when leaving the LTE coverage.

- *OTT (Over-the-Top) Applications:* Services received over the Internet that are not directly provided by the Internet Service Provider (ISP).

Additional information about VoLTE is provided in Chapter 1 of this book.

2.3.3 LTE-Advanced

LTE-Advanced or LTE-A[34] refers to specifications in 3GPP Releases 10 and beyond, focusing on higher capacity or throughput and efficiency. To this end, the main new functionalities introduced in LTE-A are as follows:

- Wider radio channels, up to 20 MHz.
  - Possible channel bandwidths include 1.4, 3, 5, 10, 15, or 20 MHz.
- Carrier aggregation,[35] up to 100 MHz.

- It may be used for both FDD and TDD. A maximum of five component carriers (the aggregated carriers) can be aggregated, hence the 100 MHz maximum bandwidth. The number of aggregated carriers can be different in the downlink and the uplink, although the number of uplink component carriers is never larger than that of downlink. The individual component carriers can have different bandwidths.

- *Enhanced use of multi-antenna (MIMO) techniques:* Higher order (8×8 in the downkink and 4×4 in the uplink), multiuser, and higher mobility.

- *Heterogeneous networks or het-nets*[36]: Macrocells combined with smaller cells (microcells, picocells, or femtocells).

- Support for relay nodes (RN).
  - The possibility for efficient heterogeneous network planning (the het-nets) is increased with the introduction of RN, which are low-power base stations that provide enhanced coverage and capacity at cell edges and hot spot areas, and they may also be used to connect to remote areas without fiber connection.

- More intelligent and seamless offload.

- Coordinated Multi Point operation (CoMP).
  - Introduced to improve network performance at cell edges. In CoMP, a number of transmit points provide coordinated transmission in the downlink, and a number of receive points provide coordinated reception in the uplink.

Among the main achievements to date are the following:

- Increased peak data rate, up to 3 Gbps in the downlink and 1.5 Gbps in the uplink (Release 11).

- Higher spectral efficiency, from a maximum of 16 bps/Hz in Release 8 to 30 bps/Hz in Release 10.

- Increased number of simultaneously active subscribers.

- Improved performance at cell edges, for example, for downlink with 2×2 MIMO at least 2.40 bps/Hz/cell.

Among the improvements achieved by the last release (Release 12, 13 is currently being work on) are the following:

- *Machine-type communication (MTC):* The huge growth expected in MTC and the Internet of Things (IoT) in the near future may result in very high network signaling and also capacity issues. As a solution, a new User Equipment (UE) category is defined for optimized MTC operations.

- *WiFi integration with LTE:* It will provide operators with more control on managing WiFi sessions. Mechanisms are specified for steering traffic and network selection between LTE and WiFi.

- *LTE in unlicensed spectrum:* Offering benefits to operators such as an increase in network capacity, load, and performance.

2.4 WiFi AC (GIGABIT)

Since 2013, IEEE 802.11ac[37,38] is available as the first WiFi connection breaking the Gbps barrier. The 802.11ac standard enhances the previous IEEE 802.11n standard avoiding the 2.4 GHz band congestion by using the 5 GHz band, providing data rates of up to 6 Gbps, thanks to a channel available bandwidth of up to 160 MHz, up to eight MIMO (Multiple Input Multiple Output) streams, up to four MU-MIMO (Multi-User MIMO) for the downstream, and a 256 QAM superdense modulation.

SU-MIMO (Single-User MIMO) is already specified in 802.11n, where the access point (AP) and the station use spatial multiplexing that employ multiple antennas to transmit data in parallel. However, spatial multiplexing is not possible for mobile and other terminals having few antennas, so spatial multiplexing capabilities of APs may be underutilized. In MU-MIMO, the AP transmits simultaneously to up to four stations using a transmit beam-forming technique, a directional transmission that cancels interuser interference. This is the basis of the SDMA (space-division Multiple Access) technique.

The final data rate will depend on the MU-MIMO configuration. For example, a one-antenna AP together with a one-antenna station using a 80 MHz bandwidth yields a 433 Mbps data rate at physical link level, typical for a handheld terminal. Whether doubling the number of antennas for both the AP and the station or doubling the bandwidth, the data rate will also be doubled to 867 Mbps. On the other extreme case, an eight-antenna AP transmitting to four two-antenna stations using a 160 MHz bandwidth would provide 1.69 Gbps to each station and an aggregate capacity of 6.77 Gbps, which could ideally be used for smart TVs or laptops applications. However, these are theoretical figures since in practice the client is receiving its signal distorted by interference from the signals intended for other users, which makes the highest constellations such as 256QAM infeasible within an MU-MIMO transmission.

Computing, mobile, and consumer electronics markets will benefit from the advantages of 802.11ac, which also has an energy consumption several times lower compared to its 802.11n predecessor.

2.5 UNIVERSAL ACCESS

Universal Internet broadband access is a priority for most governments. As of today, more than half the world population do not have access to Internet (there is an estimate that half of the world population live in just 1% of the land). Nearly 1.6 billion people live in areas without mobile broadband coverage. About 3.7 billion are not subscribers to mobile broadband despite having coverage, who are either using 2G or not subscribing to mobile Internet at all[1].

Traditional telecommunications infrastructure based on terrestrial equipment cannot reach all remote locations. The advent of mobile data was a great leap forward since fixed broadband coverage is quite limited. However, the mobile network also relies on fixed infrastructure, such as mobile base stations. Operators cannot be very profitable in areas with very little population.

A typical alternative is the satellite Internet. A few geostationary satellites can cover the whole Earth, but its data capacity is limited, and the terrestrial antenna (dish) and receiver are relatively expensive. However, satellite technology is also making good progress. An example is the recent announcement (by ViaSat) of satellites with 1 *Tbps* overall network capacity (claiming to offer the same capacity as all—400—the currently existing commercial satellites together), enabling 100+ Mbps satellite Internet for residential customers.[39] Satellites are planned starting in 2019, with global coverage through three of them (one for Americas; one for Europe, Middle East and Africa; one for Asia). Moreover, satellite can be complemented with managed WiFi hotspots, in order to reduce the reception costs.

There is also commercial offering from Medium Earth Orbit satellites (approximately 8000 km above the Earth), like the one from O3b networks that launched its first satellites in 2013, with an initial fund raising of $1.2 billion. Full commercial operations were started in 2014, and currently there are 12 satellites (with a capacity around *100 Gbps*). The company has recently raised almost half billion US dollars in order to increase the constellation to 20.[40]

There are also plans for much larger constellations. One example is OneWeb that intends to launch 900 Low Earth Orbit satellites (or micro ones), with launches starting from 2018, and for which in 2016 a joint venture with Airbus (Defense and Space) has been established.[41] It raised $500 million in 2015.[42] Capacity forecast is more than *10 terabits per second*.

Another plan that is considered more speculative is the launch of 4000 satellites (750 miles above the earth), backed by SpaceX (Elon Musk), with funding of $1 billion from Google among others.[43]

Other experimental alternatives are based on flying objects that provide Internet connectivity to wide terrestrial zones. One example is balloons by Google (which fly in the stratosphere about 20 km away from the Earth). The company has signed deals in countries like Indonesia (fourth largest in the world by population,[44] widely spread in an archipelago of around 17,000 islands) in partnership with three of the four largest mobile networks operators (Telkomsel, XL Axiata, and Indosat),[45] Sri Lanka (where penetration is very low, and where the government is involved in the initiative),[46] as well as plans for India (second largest country by population, and where the government has finally approved the experiments).[47,48]

Finally, Facebook is also very keen on providing Internet access to every person in the world.[49] In order to achieve that ambition, they are working on drones (named Aquila) that can provide Internet connectivity, by staying aloft for several months. First tests are happening in 2016. Lasers are also being investigated to deliver the data from the drones. The intention is to launch 10,000 drones.

ACRONYMS

3GPP	3rd Generation Partnership Project
ADSL	Asymmetric Digital Subscriber Line
AP	access point

APON	ATM-based PON
ASON	Automatically Switched Optical Network
ATM	Asynchronous Transfer Mode
AWG	Arrayed Wavelength Grating
BPON	Broadband PON
CAPEX	CAPital EXpenditure
CSFB	Circuit Switched Fall Back
DBA	Dynamic Bandwidth Allocation
DMT	Discrete MultiTone
DOCSIS	Data Over Cable Service Interface Specification
DSLAM	Digital Subscriber Line Access Multiplexer
DSP	Digital Signal Processor
EPON	Ethernet PON
eUTRAN	Enhanced UMTS Terrestrial Radio Access Network
FDD	frequency-division duplexing
FDM	frequency-division multiplexing
FEC	Forward Error Correction
FFT	Fast Fourier transform
FTTB	Fiber-to-the-Building
FTTC	Fiber-to-the Curb
FTTH	Fiber-to-the-Home
FTTN	Fiber-to-the-Node or Neighborhood
FTTx	Fiber-to-the x
GEM	Generic Encapsulation Method
GE-PON	Gigabit Ethernet PON
GPON	Gigabit PON
GPRS	General Packet Radio Service
HDTV	High Definition TV
HFC	Hybrid Fiber-Coaxial
IEEE	Institute of Electrical and Electronics Engineers
IFFT	inverse fast Fourier transform
IoT	Internet of things
IPTV	IP television
ISDN	Integrated Services Digital Network
ITU-T	International Telecommunication Union-Telecommunication Standardization Sector

LTE	Long-Term Evolution
LTE-A	LTE-Advanced
MDU	multidwelling units
MIMO	multiple input multiple output
MTC	machine-type communication
MU-MIMO	multi-user MIMO
NG-PON	next-generation PON
OAM	Operation Administration and Maintenance
ODN	Optical Distribution Network
OFDM	orthogonal frequency-division multiplexing
OFDMA	orthogonal frequency-division multiple access
OLT	Optical Line Terminal
ONT	Optical Networking Terminals
OPEX	OPerational Expenditure
OTT	over-the-top
PON	Passive Optical Networks
PSTN	Public Switched Telephone Network
QAM	Quadrature Amplitude Modulation
QoS	Quality of Service
RF	Radiofrequency
SC-FDMA	single carrier–frequency-division multiple access
SDMA	space-division multiple access
SNR	signal-to-noise ratio
SRVCC	single radio voice call continuity
STB	Set top box
SU-MIMO	single-user MIMO
SVLTE	simultaneous voice and data LTE
TDD	time-division duplexing
TDM	time-division multiplexing
TDMA	time-division multiple access
TWDM	time wavelength-division multiplexing
VDSL	Very-High-Bit-Rate Digital Subscriber Line
VDSL2	Very-High-Bit-Rate Digital Subscriber Line 2
VoLTE	Voice over LTE
WDM	wavelength-division multiplexing
WiFi	wireless fidelity
xDSL	x Digital Subscriber Line

NOTES

1. *Infographic:* Unique mobile Internet subscribers: The impact of mobile in bringing people online. GSMA Intelligence. https://gsmaintelligence.com/research/2016/02/infographic-unique-mobile-internet-subscribers-the-impact-of-mobile-in-bringing-people-online/551/ (February 22, 2016).
2. Definitive data and analysis for the mobile industry (*Infographic:* Unique subscribers: Understanding the true reach of mobile). GSMA Intelligence. https://gsmaintelligence.com/, (https://gsmaintelligence.com/research/2016/02/infographic-unique-subscribers-understanding-the-true-reach-of-mobile/550/) (last accessed February 27, 2016 (February 22, 2016).
3. Smartphones dominating global mobile connections base. GSMA Intelligence. https://gsmaintelligence.com/research/2016/02/smartphones-dominating-global-mobile-connections-base/540/ (February 16, 2016).
4. GSA confirms 1 in 8 mobile subs using LTE in Q3 2015. Global Mobile Suppliers Association (Ovum). http://gsacom.com/paper/gsa-confirms-1-in-8-mobile-subs-using-lte-in-q3-2015/ (December 7, 2015).
5. Full Duplex DOCSIS® 3.1 Technology: Raising the Ante with Symmetric Gigabit Service. CableLabs. www.cablelabs.com/full-duplex-docsis-3-1-technology-raising-the-ante-with-symmetric-gigabit-service/ (February 18, 2016).
6. 10 Gbps at $399 per month. US Internet. http://fiber.usinternet.com/fiber-in-the-news/ http://fiber.usinternet.com/plans-and-prices/ (December 2014).
7. Croatia, Germany, and Poland join the FTTH Ranking. Fibre to the Home Council Europe (iDate). www.ftthconference.eu/images/Banners/Conference2016/News/PR20160217_FTTHranking_panorama_award.pdf (February 17, 2016).
8. Telstra launches world's first 600 Mbps-capable category 11 device. Telstra. http://exchange.telstra.com.au/2015/09/16/telstra-launches-worlds-first-600mbps-capable-category-11-device/ (September 2015).
9. Digital agenda: broadband speeds increasing but Europe must do more. European Commission. http://europa.eu/rapid/press-release_IP-10-1602_en.htm?locale=en (November 25, 2010).
10. Impact of broadband on the economy. International Telecommunications Union. www.itu.int/ITU-D/treg/broadband/ITU-BB-Reports_Impact-of-Broadband-on-the-Economy.pdf (April 2012).
11. ETSI, xDSL. www.etsi.org/technologies-clusters/technologies/fixed-line-access/xdsl (last access March 18, 2016).
12. ITU-T G.Sup50 : Overview of digital subscriber line recommendations. www.itu.int/rec/T-REC-G.Sup50-201109-I/en (last access March 18, 2016).
13. Broadband Forum, DSL Technology Evolution. www.broadband-forum.org/downloads/About_DSL.pdf
14. Gary C. Kessler, *ISDN: Concepts, Facilities, and Services*, Mcgraw-Hill Computer Communications Series, McGraw-Hill Inc., ISBN-13: 978-0070344372, 1998.
15. Timothy J. Brophy, Steven Condra, Martin Mattingly, Ron Hranac, Leonard Ra, *FTTH Evolution of HFC Plants*, Cisco, 2011 FTTH Conference and Expo, Orlando, FL, 2011. www.cisco.com/c/dam/en/us/solutions/collateral/service-provider/cable-access-solutions/ftth_evolution_hfc_plants_brophy.pdf
16. FTTH Fiber-to-the-Home Council Americas www.ftthcouncil.org/ (last accessed March 15, 2016).
17. Gigabit-capable passive optical networks (GPON): Reach extension. Series G: Transmission Systems and Media, Digital Systems and Networks. Digital sections and digital line system—Optical line systems for local and access networks. ITU-T. G.984.6. 2008. www.itu.int/rec/T-REC-G.984.6-200803-I/en.
18. 3GPP. www.3gpp.org/.
19. The evolution of mobile technologies: 1G->2G->3G->4G LTE. Qualcomm, 2014. www.qualcomm.com/media/documents/files/the-evolution-of-mobile-technologies-1g-to-2g-to-3g-to-4g-lte.pdf.
20. DongBack Seo, *Evolution and Standardization of Mobile Communications Technology*, University of Groningen, The Netherlands, and Hansung University, South Korea, ISBN13: 9781466640740, 2013.
21. Mischa Schwartz, *Mobile Wireless Communications*, Cambridge University Press, ISBN 0521843472, 2005.

22. LTE overview. www.3gpp.org/technologies/keywords-acronyms/98-lte.

23. Jon Kenneke, *Analog Cellular Telephone Technology: An Historical Reference: An Overview and Hacking Information on the Old NAMPS System*, 2nd edition, Kenneke LLC, 2011.

24. GPRS & EDGE. www.3gpp.org/technologies/keywords-acronyms/102-gprs-edge.

25. WCDMA. www.3gpp.org/technologies/keywords-acronyms/104-w-cdma.

26. HSPA. www.3gpp.org/technologies/keywords-acronyms/99-hspa.

27. UMTS. www.3gpp.org/technologies/keywords-acronyms/103-umts.

28. George Tsoulos, *MIMO System Technology for Wireless Communications*, ISBN 9780849341908, CRC Press, 2006.

29. Tao Jiang, Lingyang Song, Yan Zhang, *Orthogonal Frequency Division Multiple Access Fundamentals and Applications*, ISBN 9781420088243, CRC Press, 2010.

30. Fathi E. Abd El-Samie, Faisal S. Al-kamali, Azzam Y. Al-nahari, Moawad I. Dessouky, *SC-FDMA for Mobile Communications*, ISBN 9781466510715, CRC Press, 2013.

31. Hyung G. Myung, David J. Goodman, *Single Carrier FDMA: A New Air Interface for Long Term Evolution*, ISBN: 9780470758700, Wiley, 2008.

32. Magedanz, T., Future Internet impacts on the evolution of next generation network infrastructures and services. ITU-T Kaleidoscope, Capetown, South Africa, 2011. www.itu.int/ITU-T/uni/kaleidoscope/2011/ngn-tutorial.html.

33. Mudigonda, S., *VoLTE or VoIP over LTE—who is the ultimate winner?* IEEE Santa Clara Valley Consumer Electronics Society. 2013. www.gsma.com/network2020/wp-content/uploads/2014/03/Imagination-VoLTE-vs-VoIP-over-LTE-26MAR2013-1.pdf.

34. LTE-Advanced, 3GPP. www.3gpp.org/technologies/keywords-acronyms/97-lte-advanced.

35. Carrier Aggregation explained. www.3gpp.org/technologies/keywords-acronyms/101-carrier-aggregation-explained.

36. HetNet/Small Cells. www.3gpp.org/hetnet.

37. 802.11ac: The Fifth Generation of Wi-Fi Technical White Paper. www.cisco.com/c/en/us/products/collateral/wireless/aironet-3600-series/white_paper_c11-713103.html#_Toc383047848.

38. The WiFi evolution: an integral part of the wireless landscape. Qualcomm. www.qualcomm.com/media/documents/files/the-wi-fi-evolution-an-integral-part-of-the-wireless-landscape.pdf.

39. First global broadband communications platform to deliver affordable, high-speed Internet connectivity and video streaming to all. Viasat.www.viasat.com/news/first-global-broadband-communications-platform-deliver-affordable-high-speed-internet#sthash.x4drni3E.dpuf (February 9, 2016).

40. O3b Networks announces the closing of $460M in financing to expand its constellation and support unprecedented customer growth (O3b Company Overview, *Our Technology at a Glance*). O3b Networks, Decenber 10, 2015. www.o3bnetworks.com/o3b-networks-announces-closing-460m-financing-expand-constellation-support-unprecedented-customer-growth/ (www.o3bnetworks.com/our-story/, www.o3bnetworks.com/technology/) (last access February 27, 2016).

41. Airbus Defence and Space and OneWeb create OneWeb Satellites company—the next stage of the OneWeb adventure. Airbus Defence and Space. https://airbusdefenceandspace.com/newsroom/news-and-features/airbus-defence-and-space-and-oneweb-create-oneweb-satellites-company-the-next-stage-of-the-oneweb-adventure/ (January 26, 2016).

42. Oneweb announces $500 million of a round funding with group of leading international companies. OneWeb. http://oneweb.world/press-releases/2015/oneweb-announces-500-million-of-a-round-funding-with-group-of-leading-international-companies (January 26, 2016).

43. SpaceX is backpedaling on its plan to launch 4000 satellites into space. *Business Insider*. www.businessinsider.com.au/elon-musks-spacex-is-backpedaling-its-satellite-internet-plan-2015-10 (October 29, 2015).

44. List of countries and dependencies by population. Wikipedia.https://en.wikipedia.org/wiki/List_of_countries_and_dependencies_by_population (last access February 27, 2016).

45. Bringing the Internet within reach of 100 million Indonesians. Google.https://googleblog.blogspot.com.es/2015/10/indonesia-loon-internet.html (October 28, 2015).

46. Google's 'Project Loon' balloon Internet experiment floats into sri lanka. *Wall Street Journal*. http://blogs.wsj.com/indiarealtime/2016/02/16/googles-project-loon-balloon-internet-experiment-floats-into-in-sri-lanka/ (February 12, 2016).

47. Google CEO Sundar Pichai: We will bring Project Loon to India soon. *The Times of India.* http://timesofindia.indiatimes.com/tech/tech-news/Google-CEO-Sundar-Pichai-We-will-bring-Project-Loon-to-India-soon/articleshow/50203190.cms (December 16, 2015).
48. Indian government gives green light to project loon expansion. Android Headlines. www.androidheadlines.com/2016/01/indian-government-gives-green-light-to-project-loon-expansion.html (January 27, 2016).
49. Inside Facebook's ambitious plan to connect the whole world. Wired.. www.wired.com/2016/01/facebook-zuckerberg-internet-org/ (January 19, 2016).

Chapter 3

Convergence: Bundling Fixed Line and Mobile Services

Integrated operators offer to customers bundles with fixed line and mobile services described in the previous two chapters: fixed line voice, fixed line broadband, and mobile services (that include voice and broadband). In recent years, operators have started to offer discounts for these bundles. This has triggered a change in the competitive landscape, with industry transitioning to convergent operators, enabled by in-country consolidation of formerly fixed line-only (e.g., cable operators) and mobile-only operators (e.g., Vodafone acquiring local operators in Europe, BT acquiring mobile operator in the United Kingdom), as well as mergers of formerly fixed line and mobile divisions of the same operator (e.g., Verizon in the United States). This trend has only started, and there are still many mobile-only players that are rethinking their strategy. Fixed line-only ones are fewer, and can complement their fixed line assets by setting up Mobile Virtual Network Operators.

3.1 CUSTOMER NEED: ONE-STOP SHOP

Arguably, customers prefer to have one single provider for all their telecommunications (and more arguably digital) needs. It is what we might call a one-stop shop that caters in a unified way all the services together. Advantages could be to have just one installation, same support center if any problem arises, a single bill (if it is not shocking), and so on. Having said all this, probably the very main advantage is that of price, since typically convergent offers have significant discounts (compared to the very same services purchased independently), which are attractive for customers.

3.2 FIXED LINE AND MOBILE SERVICE BUNDLES

In the old times, customers started buying fixed line phone (telephony) service. Pay TV was also available, through broadcast technologies such as satellite and cable.

Digital Services in the 21st Century: A Strategic and Business Perspective, First Edition.
Antonio Sánchez and Belén Carro.
© 2017 by The Institute of Electrical and Electronics Engineers, Inc. Published 2017 by John Wiley & Sons, Inc.

As technology made progress, new services appeared, mobile telephony, fixed line Internet (dial-up) that afterwards became fixed line broadband, and more recently mobile broadband. As of today, it is quite typical that telecommunications customers subscribe to a convergent package, also known as a bundle, which comprises all these services: fixed line phone, fixed line broadband, mobile phone and broadband, and optionally Pay TV. This is referred to as quadruple play. It could be said that among them, Pay TV is the one that is less massive, or in other words, the one with less adoption (as a matter of fact premium TV is an add-on, although free basic TV can be included in the bundle), but this depends on the market. Other attempts to include additional services have not been that successful so far.

One of the first telecommunications operators offering convergence was *Movistar (Telefonica)* in Spain with the name *Fusion*, authorized by the national regulator in September 2012. In 2015, it had reached 4.2 million customers, compared to less than 6 million fixed broadband customers, therefore representing more than two thirds of them. It has to be noted that the operator has 10 million fixed lines, which means approximately 4 million fixed lines do not have broadband, and therefore are not attracted by the convergence bundle. It also has 1.5 million mobile lines add-ons, typically more than one mobile line linked to each convergence bundle. However, the number of mobile lines—excluding terminals—is almost 15.5 million, including almost 3 million prepay customers for which there is no convergence offer. This shows that there is still a large part of the market that is not bundled. The average revenue per user has varied throughout the time, depending on the promotions and the amount of voice minutes and broadband gigabytes included, and was almost €73 in 2015. The churn stood at 1.2%, compared to 1.6% as in the pure mobile business.[1]

Another example in Europe is Deutsche Telekom in Germany. Its integrated offer comprising fixed line, mobile, and TV, called *Magenta Eins*in Germany, *Magenta One* internationally, was launched in the fall of 2014.[2] It reached almost 3 million customers in Europe after being launched in six markets, of which around 2 million are in Germany, where there is a target to reach 3 million by 2018. As of 2016, there are three options for the bundle: The low range costs €54.90 with 500 MB mobile data, and up to 16 Mbps for fixed line broadband, the mid-range costs €69.85, which increases to a fixed line broadband speed of up to 50 Mbps and includes TV, and the high range for €79.85 with 100 Mbps plus premium TV.

3.3 INTEGRATED OPERATORS

In the old times, when mobile voice was emerging as a mainstream services, there were several telecommunications operators that split their mobile division and in several cases made it a public company (although retaining a majority stake). Nowadays there is only one major telecommunications operator that maintains this split, namely, NTT in Japan, whose mobile subsidiary is NTT Docomo (worth 10.778 trillion yen,[3] with 4 billion outstanding shares). The parent NTT itself is worth $93.143 billion, which is roughly by chance similar in size as its subsidiary, making the two companies the fourth and fifth largest telecommunications operators by market cap in the world. NTT owns approximately two thirds of NTT Docomo as of September 30, 2015, and 66.65% of the voting rights.[4]

Most other telecommunications operators worldwide keep both fixed and mobile units together. In other words, there is not much of a distinction between them any longer. In some cases, multinational operators have made initial public offering (IPO) for their selected local operations (such as Telefonica in Brazil and Germany) in order to raise funds.

The Chinese market went through a consolidation process in 2008. The Chinese operators are run by the state, which is the main owner, through its Ministry of Industry and Information Technology (MIIT), which is also responsible for awarding 3G and 4G licenses. The consolidation resulted in three major operators that combined the fixed and mobile assets of the country. China Unicom merged with China Netcom (fixed line telecommunications)[5] while at the same time divesting its CDMA mobile subscribers to China Telecom (another fixed line telecommunications provider).[6] China Mobile had in turn been divested from China Telecom before 2000 (1999), and in 2008 acquired another fixed line telecommunications operator (China Tietong, which had been established by the Ministry of Railways). In turn, China Telecom's fixed assets had been split around 2002 (between north— the smaller part that became Netcom—and south).

Another major market that experiences convergence consolidation is that of the United States. In 2014, Verizon completed the acquisition of the 45% stake that Vodafone had in Verizon Wireless, which was valued at $130 billion.[7] The company had been established back in 2000, as a joint venture between the two companies. After the transaction, Verizon integrated its former wireless division with the parent fixed line company, becoming one of the largest telecommunications operators worldwide, together with AT&T (also from the United States) and China Mobile (from China, mentioned above). As of the end of the first quarter of 2016, each of them is roughly worth $225 billion (market cap), far away from the fourth (NTT also mentioned above, worth around $100 billion, order of magnitude).

Vodafone used to be the largest mobile-only telecommunications operator in the world. However, convergence triggered the transition of its profile, with inorganic growth making it become an integrated operator in several countries. In all three cases, the combination has been with a cable operator. In the first two cases, an acquisition, whereas in the third case a joint venture. In 2013 it acquired Kabel Deutschland, with an equity value of approximately €7.7 billion (enterprise value approximately €10.7 billion including debt),[8] representing a premium of 37% compared to the share price before takeover speculation started on February 12, 2013 and around four times more compared to the IPO 3 years before (March 2010). Similarly, in Spain in 2014, it acquired a cable operator for a similar price, about €7.2 billion (valuation at multiple of 7.5× 2013 EBITDA).[9] By then, it was serving 1.9 million customers (and offered coverage to more than 7 million households, equivalent to approximately 40% of the total). Again, it strengthened its unified communications strategy. Most recently, Vodafone agreed to establish a joint venture in The Netherlands with Liberty Global (local brand Ziggo), the largest multinational cable operator in the world (with operations in 14 countries)[10] with total synergies for an estimated net present value of €3.5 billion. Vodafone will make a cash payment of €1 billion to Liberty Global in order to have 50% ownership of the merged business (based on enterprise value and after deducting €7.3 billion of net debt of Ziggo). This transaction was referred to by Vodafone as a continuation of their convergence strategy market by market.

Another large European market witnessing the convergence trend is United Kingdom. In 2015, BT (a fixed line-only telecommunications operator, after divesting its mobile division O2 several years ago) announced the agreement to acquire mobile operator Everything Everywhere (jointly owned by Orange and Deutsche Telekom, who had in the past merged their mobile business in UK market) for £12.5 billion.[11] In January 2016, the competition regulator (*Competition and Markets Authority's*) approved the acquisition unconditionally without remedies.[12]

Another region with convergence consolidation is Latin America. Particularly, Brazil has witnessed two consecutive M&A operations from Telefonica. In 2010, Telefonica acquired Portugal Telecom's 30% stake in mobile telecommunications operator Vivo for €7.5 billion (all cash), valuing the company at a multiple of 10 times its EBITDA.[13] Telefonica already owned 30%, with the remainder floating publicly, for which a partial offer was made for €0.8 billion. After closing the operation, it merged the company with its fixed telecommunications subsidiary (*Telecomunicacoes de Sao Paulo-Telesp*). Five years later, in March 2015, Telefonica acquired the fixed telecommunication operator in Brazil GVT (Global Village Telecom), owned by Vivendi (which had bought it a few years earlier, precisely in competition with Telefonica). Its founder and CEO became CEO of the new combined company.[14] The new combined entity became the largest operator in the market in terms of both customers (accesses) and revenues. The price included cash, €4.67 billion, and shares, 12% of the capital of Telefónica Brazil, after the integration. Also the parties agreed to exchange 4.5% of the integrated company by Telefonica's share in an Italian operator. The main reason behind this acquisition was geographical, in the sense that the acquirer used to be the incumbent fixed line operator in a state of the country (Sao Paulo) but lacked fixed line assets in most of the remainder of the country, where the acquired company had stronger assets.

Another example of integration is where a telecommunications operator does have fixed line and mobile assets, but similar to the previous case its competitive positioning is weaker in one of them. That was the case of Orange (France Telecom) in Spain. Originally, it was a small fixed operator trying to compete with the incumbent. In 2005, it acquired an 80% stake of a mobile network operator (an acquisition that was authorized by European Commission[15]) for €6.4 billion.[16] Long after that, it had the intention to reinforce its fixed line assets, following the convergence offer launched by leading player, and the consolidation of mobile and fixed competitors mentioned earlier. In 2014, it acquired fixed line operator Jazztel, which also had mobile customers through an MVNO, for an enterprise value of €3.8 billion (with a valuation multiple of 8.6 times EBITDA), an all cash offer with a premium of 34%, compared to average trading price over the previous 30 days.[17]

ACRONYMS

CDMA   Code Division Multiple Access
EBITDA   Earnings Before Interests Taxes Depreciation and Amortization
IPO   Initial Public Offering
M&A   Mergers and Acquisitions
MB   megabyte
MVNO   Mobile Virtual Network Operators
TV   television

NOTES

1. January–December 2015 Results. Telefonica, //www.telefonica.com/en/web/shareholders-investors/financial_reports/quarterly-reports (February 26, 2016).
2. Results for the 2015 financial year. Deutsche Telekom, //www.telekom.com/static/-/298764/9/160225-q4-allinone-si (February 25, 2016).
3. NTT Docomo Inc./Nippon Telegraph and Telephone Corp. Bloomberg, //www.bloomberg.com/quote/9437:JP //www.bloomberg.com/quote/NTT:US (last accessed March 27, 2016).
4. Corporate Governance. NTT Docomo. //www.nttdocomo.co.jp/english/corporate/ir/management/governance/#p04 (last accessed March 27, 2016).
5. China Unicom dispatches circular on merger with China Netcom—aiming to become a world-class broadband communications and information services operator. China Unicom. //www.chinaunicom.com.hk/en/press/press_content.php?id=p080814 (August 14, 2008).
6. Milestones. China Telecom. http://en.chinatelecom.com.cn/corp/Highlights2008/index.html (last accessed March 27, 2016).
7. Verizon completes acquisition of Vodafone's 45% indirect interest in Verizon Wireless. Verizon. //www.verizon.com/about/news/verizon-completes-acquisition-vodafones-45-percent-indirect-interest-verizon-wireless (February 21, 2014).
8. Kabel Deutschland welcomes announced tender offer from Vodafone. Kabel Deutschland. //www.kabeldeutschland.com/static-com/tx_kdgnews/20130624_KDH_PM_Vod_Offer_final.pdf (July 24, 2013).
9. Vodafone to Acquire Grupo Corporativo Ono, S.A. Vodafone. //www.vodafone.com/content/dam/vodafone-images/investors/pdfs/acquisition-of-ono-newsrelease.pdf (March 17, 2014).
10. Liberty Global and Vodafone to merge their Dutch operations. Vodafone. //www.vodafone.com/content/dam/vodafone-images/investors/pdfs/160216_lg_vf_merge_dutch_ops.pdf (February 16, 2016).
11. General meeting to approve the proposed acquisition of EE. BT. //www.btplc.com/Sharesandperformance/AGMs/Generalmeeting2015/index.htm (April 30, 2016).
12. BT Welcomes CMA'S Approval of EE Acquisition. BT. http://btplc.com/Sharesandperformance/Presentations/Presentations/keycompanyannouncements/downloads/DC16-018BTwelcomesCMAsapprovalofEEAcquisition.pdf (January 15, 2016).
13. Telefonica agrees to buy Portugal Telecom Vivo stake. Bloomberg. //www.bloomberg.com/news/articles/2010-07-28/telefonica-said-to-agree-9-8-billion-acquisition-of-vivo-stake-in-brazil (July 28, 2010).

14. Telefónica buys GVT and becomes undisputed leader in the Brazilian market. Telefonica. //www
.telefonica.com/en/web/press-office/news-detail/-/asset_publisher/h6pNStBlfIXK/content/telefonica-
buys-gvt-and-becomes-undisputed-leader-in-the-brazilian-market (March 25, 2015).

15. Case No. COMP/M.3920—France Telecom/Amena. Notification of 19/09/2005 pursuant to Article 4
of Council Regulation No 139/20041. European Commision. http://ec.europa.eu/competition/
mergers/cases/decisions/m3920_20051024_20310_en.pdf (October 24, 2016).

16. France Telecom buys Spain's Amena. BBC. http://news.bbc.co.uk/2/hi/business/4720163.stm (July
27, 2005).

17. Acquisition of Jazztel; creating a key convergent player in Spain. Orange. //www.orange.com/fr/
content/download/25384/573037/version/10/file/Orange+acquires+Jazztel+-+Investor+Presentation
+EN+-+vDEF2+-+disclaimer.pdf (September 16, 2014).

Chapter 4

Devices: Smartphones

The launch of iPhone in mid-2007 had a big impact on telecom industry, accelerating the take-up of mobile broadband and disrupting mobile device industry (with formerly dominant players such as Nokia, Blackberry, and Motorola failing to respond and becoming almost insignificant). In 2015, 1.4 billion smartphones were sold (14% more than in 2014, when smartphone represented two thirds of global phone sales, with 1.2 billion shipments).[1] Samsung and Apple lead the market with double digit share, followed by Chinese vendors, but this is only an illustrative snapshot, because every quarter vendors compete fiercely for sales. Lower smartphone prices have accelerated the migration from earlier feature phones.

The clear dominant Operating System is Android, having surpassed 80% market share in the fourth quarter of 2015, while iOS from Apple devices has almost 18%. Challengers remain marginal, almost negligible with Windows Phone just accounting for 1% (with decreasing trend), and all others (including Blackberry) accounting for less than 0.5%.

Regarding technical features, the trend is to have larger screens (e.g., Huawei P8max with almost 7 in. diagonal), and thinner and ideally less weight. Besides physical features, battery duration is also important for customers. Specifications vary in connectivity capabilities (as shown in Chapter 2), processor speed, storage capacity, cameras resolution, sensors, and so on. Of course, price and design are also key decision factors. However, 9 years after the first smartphone, it seems that it is increasingly difficult to innovate in the space, as can be seen in the evolution of releases of flagships/best sellers: Apple iPhone (latest 6s) and Samsung Galaxy S (latest 7).

4.1 CUSTOMER NEED: MOBILITY

Before the arrival of smartphone, there was a clear distinction between a personal computer (more specifically a laptop) and a mobile phone (the so-called feature phone). Both fill the customer need of mobility. The former covered the same

Digital Services in the 21st Century: A Strategic and Business Perspective, First Edition.
Antonio Sánchez and Belén Carro.

computing functions of a desktop, whereas the latter was mainly used to make phone calls (with a few additional features such as text messaging—SMS, and entertainment). However, a laptop was too heavy and too large to be carried around in an *effortless* manner.

The appearance of smartphone was a key success factor for the adoption of mobile Internet, since feature phones (small non-touch screens) were cumbersome to surf the Internet. Therefore, mobility (for information technologies) can be considered a key customer need addressed by smartphones.

4.2 VENDORS

As already commented in the introduction, Samsung and Apple lead the market with double digit shares, followed by Chinese vendors (Huawei, Lenovo, and Xiaomi). Apple's financial results were very impressive quarter after quarter, with a market cap above $.5 trillion in the beginning of 2016,[2] having even reached $0.75 trillion. Samsung's market share declined in 2015 compared to 2014, and Apple sales declined for the very first time in the last quarter of 2015 compared to the same quarter in 2014. These numbers are only an illustrative snapshot, because every quarter vendors compete fiercely for sales.

It is remarkable to see how former leaders Nokia, Motorola, and Blackberry have lost their position, and in some cases have been absorbed by others (e.g., Nokia by Microsoft, Motorola by Google and subsequently by Lenovo) with very low valuations compared to their peaks.

On the other hand, new vendors have emerged very strongly, as is the case of Xiaomi, which is a private company with a valuation of $45 billion after the founding round in April 2015. It is the second largest nonpublic company in the world at the beginning of 2016, having raised $1.1 billion.[3]

Apple sold in the last quarter of 2015 almost 75 million units of iPhone, its only model of smartphone, with more than $51 billion in revenues, representing more than two thirds of the total revenues of the company.[4] Both units sold and revenues were basically flat over the same quarter in the year before. It is quite clear that iPhone is the best-selling smartphone, and it is the key product behind the largest company in the world by market cap (as of early March 2016).

On the other hand, Samsung sold more than 300 million smartphones in 2015, but its average selling price is lower than that of iPhone, and it is estimated that Apple absorbs most of the profits of the smartphone industry.

Other large players are ZTE (similar to Huawei, as both are telecommunications equipment vendors, and both are based in China), HTC (smartphone vendor from Taiwan), LG (similar to Samsung, both consumer electronics behemoths from South Korea), and Sony (similar to LG but from Japan). The list is much longer with other players such as Google (for which the smartphones are manufactured by others, rather than a standalone business by itself), Asus (Taiwan), Micromax (India), Spice (India), and so on.

4.3 OPERATING SYSTEM DUOPOLY

As already commented, Android clearly dominates in the mobile device Operating Systems, having surpassed 80% market share (fourth quarter 2015). This could be almost called monopoly, but the second contender has almost the remaining 20% (iOS for Apple devices). Efforts by challengers to compete in this space have not proven very successful, with their market shares almost negligible. Only Microsoft with a large investment managed to have some relevance with Windows Phone (its predecessor was Windows Mobile), but its trend is negative and it just represents 1%. Others such as Blackberry OS, Symbian (by Nokia), and Palm OS were strong in the past but have shrunk tremendously or have been abandoned. Other attempts around Linux—even promoted by large manufacturers such as Tizen by Samsung—, or around the web—such as Firefox OS from Mozilla sponsored by telecommunications operators—have not made much progress either.

The very main reason for this situation is the very strong ecosystems that have been created around the two dominant Operating Systems. Customers only want to buy smartphones with Operating Systems that support the widest portfolio of applications (apps). On the other hand, developers (including large software companies) only focus on those Operating Systems for which there is a critical mass of customers.

Apple claims to have reached 1 billion active devices in 2015.[4] This includes non-iOS devices, such as Mac computers, whose sales are much lower, with orders of magnitude less than the installed base of iOS devices.

It is interesting to note that in both cases the Operating System is not the core business of either Apple or Google. On the one hand, there is a device manufacturer, whose revenues come mainly from the sale of hardware. On the other hand, there is an Internet player, whose revenues come predominantly from advertising (more specifically Internet advertising). In the former case, it is a vertically integrated company, encompassing hardware and software (which could be considered free); whereas in the latter case, almost all other manufacturers integrate its Android Operating System, since in most cases they do not have the required critical mass to compete in the Operating Systems space.

A key success factor for both platforms is their applications stores, called App Store (iOS) and Google Play (Android). Apple App Store[5] has paid almost $40 billion to App developers since 2008, with more than one third of the total in 2015, taking into account that developers get 70% of the revenues (same applies for Android).[6] The New Year day in 2016 set a new milestone, with customers spending almost $150 million (Apps and in-Apps purchases); contributing to a record Christmas season (two weeks ending 3rd January) surpassing $1.1 billion. In 2015, Apple's App Store sales were over $20 billion (estimated to be 75% bigger than Google Play[8]). Apple estimates to have created over 1.4 million jobs in the United States, 1.2 million in Europe, and 1.4 million in China, which are attributable to App-economy.

The number of Apps has grown considerably. It is estimated that there are 2 million Android Apps.[7] Also impressive is the number of downloads. It is estimated that Google Play has 100% more downloads than App Store.[8]

4.4 HARDWARE SPECIFICATIONS

Initial cellular terminals' appearance was similar to a cordless telephone, integrating a hard numeric keyboard and a small LCD screen oriented only to voice calls and SMS.

A smartphone differs from its predecessors in several aspects, the following being the most important:

- A large touch screen that enables high-resolution video and picture watching, gaming, reading, web browsing, email, camera viewfinder, and the visual support for all kinds of applications.
- Mobile broadband data that enables wireless Internet access via cellular networks (3G/4G) or fixed wireless networks such as WiFi.
- Very high computing capacity, as an anecdote NASA says today's "cell phones have more computing power than the computers used during the Apollo era."[9]
- Higher capacity of the battery, but with low duration mainly due to the screen and wireless data transfer's power requirements.

There are signs of smartphone market saturation in developed markets, while stronger growth has been found in developing and emerging economies.[10] This may be due to the fact that fast improvements in standard technology make consumers able to enjoy high-performance terminals even if the device is not so recent and may be two or more years old. Also, in developed markets, customers are becoming more price-sensitive and less brand-orientated, since low-cost brands, especially those coming from China, have reached the market with less expensive terminals that do not necessarily mean less quality.

Technical specifications of smartphones include the following:

- *Display:*
  - Large touch screen typically ranging from 4 to 6in. (if it is over 5 in., the terminal is commonly known as phablet, a combination of phone plus tablet, which is typically over 7 in.), measured diagonally.
  - Although there are many terminals supporting high definition (HD) with 1280×720 pixels, Full High Definition (FHD) with 1920×1080 pixels is very common. Some current top-level terminals support Quad HD with 2560×1440 pixels, and even Ultra-HD (UHD) with 3840×2160 pixels,[11] which dramatically raise their power consumption and have a questionable impact in visualization improvement in small screens. In fact, resolution is a measurement of pixel density and it depends on the screen size and the definition (number of pixels), so the smaller the display, the higher the resolution, and vice versa. For example, for FHD with a 5 in. screen, the pixel density would be around 440 ppi (pixels per inch), which drops to 400 ppi for a 5.5 in. size.
  - *Display type*[12–15]: There are two major technologies on the market, AMOLED (Active Matrix Organic Light-Emitting Diode) or Super

AMOLED (Samsung's brand[16]) and IPS (In-Plane Switching) LCD (Liquid Crystal Display).

- In AMOLED individual pixels are lit separately (active matrix: Each pixel is attached to a transistor and capacitor individually) on top of a thin film transistor (TFT) array that passes electricity through organic compounds (OLED). Some OLED screen advantages include lightweight, great responsiveness, bright and bold colors, better definition than most LCD screens, and less battery hungry in many cases since OLED displays are "always off" unless the individual pixel is electrified. Also, AMOLED displays may be thinner than LCDs and may be flexible. The disadvantages are that they age quickly and can lose color clarity for red and blue, and are more expensive than LCDs.
- LCD uses polarized light, which is then run through a color filter. Horizontal and vertical filters on either side of the liquid crystals control the brightness and whether or not each pixel is on or off. TFT LCD uses an active matrix and has lower cost but poorer image quality and has higher power consumption. IPS LCD[17] rotates the orientation of the crystal array, improving the viewing angles, contrast ratio, and color reproduction.

- *CPU (Central Processing Unit) or processor*[18]: Most smartphones are equipped with multicore processors that enable multitask and a fast terminal response and hence better performance. Current terminals are 64-bit and have up to eight cores (octa-core), typically using different types of cores for different functions: Four of the cores are used for high-performance computing and the rest have better power efficiency. The underlying OS picks the best processor core for the assigned task, for example, a background task that is just checking email does not need to use the high-performance core, but when playing a game, a high-performance core would be needed. Samsung Exynos 8890 and Qualcomm Snapdragon 810 are examples of octa-core processors. The processors' clock rate also impacts performance, with values around 2 GHz to be quite common.

- *GPU (Graphics Processing Unit):* As the name suggests, it is a processor exclusively dedicated to graphics processing or floating point operations. It alleviates the work of the CPU and dedicates itself for gaming or applications with 3D interactive graphics. Its architecture is similar to the CPU but typically including more cores and a lower clock rate. According to the inventor of the GPU,[19] its processor will enable fast multimedia, great gaming, seamless multitasking, extraordinary productivity, astounding battery life, and brilliant graphics.

- *Storage and memory:* It comprises the RAM (Random Access Memory), the internal memory, and the external storage.
  - *RAM:* It stores data or applications the terminal is using in a rapid and less power-consuming way than the internal or external storage. It is also essential to allow for background processes and quick switching between

applications without lag or delay. The way the Operating System (OS) interacts with the RAM will determine its performance: 1 GB RAM may work properly with certain OS but go slowly with a newer OS version. Current top-level Android terminals incorporate 4 GB RAM, and the first 6 GB RAM smartphone has recently been released.[20] It incorporates the Samsung 12 GB LPDDR4 (Low Power, Double Data Rate 4) mobile DRAM[21] (Dynamic RAM) that enables better UHD video recording and playback capabilities, thanks to the chip ability to move data at rates of over 4 Gbps.

- *The internal memory is composed of two parts:* One part (also called firmware) for supporting the Operating System and other files such as preinstalled applications and that in theory cannot be written by the user, and a second part destined for the user's storage, that oscillates between 80 and 90% of the overall memory capacity.
- *External storage:* Although some terminals do not allow storage expansion through an SD (Secure Digital) card, promoting the cloud storage since internal memory capacity is not enough for today's needs of storage, most smartphones are able to extend the internal memory up to 200 GB in most recent terminals, and even 2 TB in one new terminal.[22]

• *SIM (Subscriber Identity Module):* SIM size has evolved from its original size to micro-SIM and nano-SIM, the last two commonly employed nowadays. Dual-SIM terminals may use two different rate SIMs. eSIM (Embedded SIM) presupposes no external chip at all. Instead, all subscriber data are stored on the device itself, initially downloaded remotely. This will pave the way for the Internet of Things in the near future, and the first smartwatch incorporating eSIM[23] is already available on the market.

• *Mobile networks:* The network typically supports 2G, 3G, 4G, and their variants. The latest cellular network being introduced is LTE-Advanced. Frequency bands that may differ among different countries are also specified.

• *Connectivity:* Physical connectors as well as wireless connections are indicated:
- Headphones, for example, stereo 3.5 mm.
- *USB (Universal Series Bus):* Most of today's terminals incorporate USB 2.0 ports for charge and data transfer with PC. If an Android device supports USB OTG (On-The-Go), then it is possible to couple it with peripheral devices, such as a mouse or a keyboard, which can then be used to control it. USB memories can also be connected. USB type-C is already being adopted by some terminals.[24]
- *Geographic location:* Satellite location systems access such as GPS (Global Positioning System) or Glonass (Global'naya Navigatsionnaya Sputnikovaya Sistema) are common.
- *Bluetooth:* The latest version is Bluetooth 4.2., which includes Smart Bluetooth.
- *NFC* [25] *(Near Field Communication):* NFC enables simple and safe two-way interactions between electronic devices, allowing consumers to

perform contactless transactions, access digital content, and connect electronic devices with a single touch. It is based on ISO/IEC 14443 A&B and JIS-X 6319-4 standards. NFC enables the terminal to share information at a distance less than 4 cm with a maximum communication speed of 424 kbps. The user can share business cards, make transactions and payments, access information from a smart poster, or provide credentials for access control systems with a simple touch.

- *WiFi* [26]: Most terminals support IEEE 802.11a/b/g/n in the 2.4 GHz and 5 GHz bands. Recent terminals also support IEEE 802.11ac that incorporates MU-MIMO (Multi-User Multiple Input Multiple Output). WiFi Direct[27] is also an option: It enables WiFi devices to connect directly, simplifying tasks such as printing, sharing, synchronizing, and displaying. Products bearing the WiFi Direct certification mark can connect to one another without joining a traditional home, office, or hotspot network.
- *DLNA (Digital Living Network Alliance)*[28]*:* To directly share media through a common network (e.g., WiFi) with other DLNA compatible devices such as Smart TVs.

• *Sensors:* A sensor is a sensing device that measures a physical quantity and converts it into a signal. This signal can then be used by applications to offer valuable data to the user. The Android platform supports three broad categories of sensors[29]:
- Motion sensors, the most common ones are the following:
  ◦ *Accelerometer:* It measures the acceleration force in m/s^2 that is applied to a device on all three physical axes (x, y, and z), including the force of gravity.
  ◦ *Gyroscope:* It measures a device's rate of rotation in rad/s around each of the three physical axes (x, y, and z).
- Environmental sensors, the most common ones are the following:
  ◦ *Light:* It measures the ambient light level (illumination) in lux (lx).
  ◦ *Pressure:* It measures the ambient air pressure in hPa or millibar.
  ◦ *Relative humidity:* It measures the relative ambient humidity in percent (%).
  ◦ *Ambient temperature:* It measures the ambient room temperature in degrees Celsius (°C).
- Position sensors, the most common ones are the following:
  ◦ *Magnetic field:* It measures the ambient geomagnetic field for all three physical axes (x, y, z) in µTesla (µT).
  ◦ *Proximity:* It measures the proximity of an object in centimeters relative to the view screen of a device. This sensor is typically used to determine whether a handset is being held up to a person's ear.

Additional sensors, such as fingerprint sensors, enable biometric verification for secure device and website authentication as well as mobile payment. Sensors are capable of providing raw data with high precision and accuracy that would be processed by applications to offer valuable data to the user. For example, a mapping application will employ data from the magnetometer to detect the direction of magnetic north and, in conjunction with GPS, will determine the user's location. In combination with the accelerometer data it will be able to track the user's trajectory.

- *Camera:* Current terminals incorporate two cameras, one frontal and one rear. Rear cameras have traditionally been technically more advanced. However, with the boom of selfies (photograph taken with a smartphone or other digital camera by a person who is also in the photograph especially for posting on a social media website) frontal cameras are getting better and may even include flash. The resolution (number of megapixels) and video recording definition are parameters usually used for specifying a camera; however, it is not always true that more resolution translates into higher picture quality. One megapixel refers to 1 million pixels per inch. The main parameters are as follows:

  - *Lens:* Smartphone cameras usually employ fixed focal length wide angle (around a 28 mm equivalent) lenses. This means zooming in on a subject involves using image-deteriorating digital zooms. Some manufacturers offer terminals with optical zoom.[30] And it is possible to use add-on lenses[31,32] to improve final image quality results: fish-eye, telephoto, and macrolenses.
  - *Sensor size:* Light enters through the camera lens and then impinges onto the camera sensor, which receives the image information and translates it into an electronic signal. From there, the image processor creates the image and fine-tunes it to correct for a typical set of photographic flaws, such as noise. The size of the image sensor is quite important: In general, larger the sensor, larger the number of pixels, hence the large amount of light that can be collected. More light (generally) equals less-noisy images and greater dynamic range. Many sensors are CMOS (complementary metal–oxide–semiconductor) and the Sony Exmor RS™[33] stands out in the market as of today.
  - *Aperture:* Larger aperture (lower numbers, e.g., f/1.4) means better quality.
  - *Image stabilization:* Optical Image Stabilization (OIS) offers better performance than digital image stabilization.
  - *Image processing:* Quite important in the final quality of a picture. CPUs usually have dedicated graphics processors built into the chip, which, being hardware-accelerated and not just software-dependent can quickly render images such as photos, videos, and games without overtaxing the main application processor.

  Recent terminals[34] include a novel dual-camera system that promises optical zoom with no moving parts, while also allowing for better noise reduction. A combination of one monochrome and one color sensor captures three times the light of an ordinary camera, resulting in much better sharpness and clarity.

- *Battery:* Made of lithium-ion, they do not suffer from memory effect similar to the nickel–cadmium predecessors. Battery capacity is measured in mAh (milli Ampere hour). The higher the mAh rating, the larger the battery capacity that would theoretically make the battery last longer. However, other factors such as screen brightness and resolution, together with the Operating System management, also play a big part in the battery drain. Usually, the lasting time for tasks such as 3G, 4G, or WiFi Internet use, audio and video play, and 3G conversation is indicated (in hours). Batteries may be removable or nonremovable, which makes it impossible to be changed if deteriorated. External USB battery packs are employed to extend the battery duration when no charging is possible. Some terminals support wireless battery charging, avoiding the need for cabling and the deterioration of the USB port.

- *Sound:* Although built-in smartphone speakers still cannot compete against dedicated high-end portable players, some terminal options, such as an adequate placement, would increase the sound quality, for example, dual frontal stereo speakers with built-in amplifiers.[35] Some manufacturers opt to utilize the audio codec found within the phone's chipset, whereas others prefer to use dedicated chips to improve the terminal audio quality capabilities.

- *Specific Absorption Rate (SAR)*[36,37]: SAR is a measure of the rate of RF energy absorption by the body from the source being measured. The SAR limit for mobile devices is 2 W/kg. Tests for SAR are conducted using standard operating positions with the device transmitting at its highest certified power level in all tested frequency bands. Head and body SAR figures are commonly provided by terminal manufacturers.

- *Other features:*
  - – FM radio, optional since audio streaming is very popular.
  - – *Terminal resistance:*
    - ○ *Protection against display damage*[38]: Resistance to scratch and sharp contact damage, drop performance.
    - ○ *Waterproof or dustproof, specified with the IP (International Protection) rating*[39]: For example, IP65 and IP68 means the terminal is protected from dust and against low-pressure water jets, such as a faucet, when all ports are closed.[40]

As technology advances so fast, new improvements are currently being developed or researched:

- *Modularity:* A user-configured terminal is an attractive idea that could give the user freedom to choose some device configurations or modules such as battery size, camera quality, FM radio provision, or dual-screen enabling large screen temporary shut off providing a small size basic phone adequate, for example, for sport activities. The LG G5[41] is the first modular smartphone available, with two modules as of date: a camera module (LG Cam Plus), with enhanced camera capabilities and an additional battery of 1,200 mAh, and a sound module (LG Hi-Fi Plus), that converts the terminal into a Hi-Fi player.

- *Flexible bendable screens:* Though what is known as flexible screen already exists on the market in the form of curved displays, screens that are able to fold and/or blend in order to take advantage of a smaller size when not in use (e.g., for carrying it in the pocket) but with a large screen to use, are expected soon.[42,43] In fact, tablets exist mainly to provide users with a larger screen than that of smartphones for applications that do require it.

- *Improved battery duration:* Since it is one of the main drawbacks in today's smartphones, efforts are being made to extend the battery duration so that normal use might allow for more than the current one-day average duration. Graphene, the "supermaterial" discovered in 2004, is being considered to address this issue, as well as with the above mentioned bendable screens. Graphene is a fine sheet of pure carbon, as thin as an atom, the skinniest material known. At the same time, it is 100 times stronger than steel, hugely pliable and can conduct electricity and heat better than anything else. Recently Samsung[44,45] has claimed to be developing a new technology that can extend the life of a lithium-ion battery to double its current capacity on a single charge, by replacing the graphite anode, the part through which energy enters the battery, with graphene-coated silicon to make batteries with an energy density as much as 1.8 times more than that of current batteries. Researchers grew graphene cells directly on the silicon layers to allow for the expansion and contraction of the silicon and increase the battery's capacity to store two times as much energy as lithium-ion batteries with graphite anodes.

ACRONYMS

AMOLED	Active Matrix Organic Light-Emitting Diode
CMOS	complementary metal–oxide–semiconductor
CPU	Central Processing Unit
DLNA	Digital Living Network Alliance
DRAM	Dynamic RAM
eSIM	Embedded SIM
FHD	Full High Definition
Glonass	Global'naya Navigatsionnaya Sputnikovaya Sistema
GPS	Global Positioning System
GPU	Graphics Processing Unit
HD	High Definition
iOS	i Operating System (Apple)
IP	International Protection
IPS	In-Plane Switching

LCD	Liquid Crystal Display
LPDDR4	Low Power, Double Data Rate 4
mAh	milliampere hour
MU-MIMO	multiuser multiple-input multiple-output
NFC	Near Field Communication
OIS	Optical Image Stabilization
OLED	organic light-emitting diode
OTG	On-The-Go
RAM	Random Access Memory
SAR	specific absorption rate
SD	Secure Digital
SIM	Subscriber Identity Module
SMS	Short Messaging System
TFT	thin film transistor
UHD	Ultra HD
USB	Universal Series Bus

NOTES

1. Gartner says worldwide smartphone sales grew 9.7% in the fourth quarter of 2015. Gartner. //www .gartner.com/newsroom/id/3215217 (February 18, 2016).
2. Bloomberg business: Apple. Bloomberg. //www.bloomberg.com/quote/AAPL:US (last accessed February 28, 2016).
3. CrunchBase Unicorn Leaderboards. Techcrunch. http://techcrunch.com/unicorn-leaderboard/ (last accessed March 5, 2016).
4. Apple reports record first quarter results. Apple. //www.apple.com/pr/library/2016/01/26Apple-Reports-Record-First-Quarter-Results.html (January 26, 2016).
5. Record-breaking holiday season for the App Store. Apple. //www.apple.com/ca/pr/library/2016/01/06Record-Breaking-Holiday-Season-for-the-App-Store.html (January 6, 2016).
6. Apple developer programme/Google Play Developer Console Help. Apple/Google. https://developer .apple.com/programs/ / https://support.google.com/googleplay/android-developer/answer/112622? hl=en (last accessed March 6, 2016).
7. Android Statistics > Number of Android application. AppBrain. //www.appbrain.com/stats/number-of-android-apps (last accessed March 6, 2016).
8. App Annie 2015 retrospective—monetization opens new frontiers. App Annie. http://blog.appannie .com/app-annie-2015-retrospective/ (http://venturebeat.com/2016/01/20/app-annie-2015-google-play-saw-100-more-downloads-than-the-ios-app-store-but-apple-generated-75-more-revenue/) (January 20, 2016).
9. Do-it-yourself podcast: Rocket evolution, //www.nasa.gov/audience/foreducators/diypodcast/rocket-evolution-index-diy.html (last visited April 12, 2016).
10. Linda Yueh, Is the smartphone market approaching maturity? //www.bbc.com/news/business-29339251 (2014).
11. Xperia Z5 Premium, Sony Mobile, //www.sonymobile.com/es/products/phones/xperia-z5-premium/ #specifications (last visited April 13, 2016).

12. Kris Carlon, Smartphone screens explained: display types, resolutions, and more, //www.androidpit .com/smartphone-displays-explained, 2016 (last visited April 13, 2016).
13. LCD or OLED? What is the difference between smartphone screen types?, 2015, //www.amgoo .com/smartphone-blog/lcd-or-oled-what-is-the-difference-between-smartphone-screen-types (last visited April 13, 2016).
14. David Nield, Gadget tech explained: AMOLED versus IPS displays, //www.gizmag.com/amoled-vs-ips-display-technology/39196/, 2015 (last visited April 13, 2016).
15. James Rogerson, Super AMOLED versus super LCD: top smartphone screens compared, //www .techradar.com/news/phone-and-communications/mobile-phones/super-amoled-vs-super-lcd-the-big-screens-compared-1226721, 2014 (last visited April 13, 2016).
16. Samsung Galaxy S7, //www.samsung.com/us/explore/galaxy-s7-features-and-specs/ (last visited April 13, 2016).
17. iPhone 6, //www.apple.com/iphone-6/specs/ (last visited April 13, 2016).
18. Comparison of smartphone and tablet processor performance, //www.notebookcheck.net/Smartphone-Processors-Benchmark-List.149513.0.html (last visited April 13, 2016).
19. Nvidia Tegra features, //www.nvidia.co.uk/object/tegra-features-uk.html (last visited April 15, 2016).
20. Agamoni Ghosh, Vivo Xplay5 Elite: 6GB RAM smartphone introduced for first time by Chinese company, //www.ibtimes.co.uk/vivo-xplay5-elite-6gb-ram-smartphone-introduced-first-time-by-chinese-company-1547328 (2016).
21. Ian Paul, Samsung's new RAM heralds the dawn of 6GB smartphones, //www.pcworld.com/article/ 2982313/components/samsungs-new-ram-heralds-the-dawn-of-6gb-smartphones.html, 2015 (last visited April 13, 2016).
22. Discover the LG G5, //www.lg.com/us/mobile-phones/g5#G5Specs (last visited April 18, 2016).
23. Samsung Gear S2, //www.samsung.com/us/explore/gear-s2/ (last visited April 13, 2016).
24. Nexus 6P, http://consumer.huawei.com/minisite/worldwide/nexus6p/specifications.htm (last visited April 14, 2016).
25. NFC Forum, About the Technology, http://nfc-forum.org/what-is-nfc/about-the-technology/ (last visited April 14, 2016).
26. Wi-Fi Alliance, //www.wi-fi.org/ (last visited April 14, 2016).
27. Wi-Fi Direct, //www.wi-fi.org/discover-wi-fi/wi-fi-direct (last visited April 14, 2016).
28. DLNA, //www.dlna.org/ (last visited April 15, 2016).
29. Sensors Overview, http://developer.android.com/intl/es/guide/topics/sensors/sensors_overview.html (last visited April 14, 2016).
30. ASUS ZenFone Zoom (ZX551ML) //www.asus.com/es/Phone/ZenFone-Zoom-ZX551ML/.
31. The best smartphone camera accessories for 2015, //www.gizmag.com/best-smartphone-camera-photography-accessories/37703/ (2015).
32. Moment, http://momentlens.co/ (last visited April 14, 2016).
33. Sony announces the Exmor RS™, the industry's first stacked CMOS image sensor with an image plane phase detection signal processing function for high-speed AF, //www.sony.net/SonyInfo/News/ Press/201411/14-112E/ (2014).
34. Huawei P9, http://consumer.huawei.com/en/mobile-phones/p9/ (last visited April 14, 2016).
35. HTC Desire, //www.htc.com/us/smartphones/htc-desire/ (last visited April 15, 2016).
36. Specific absorption rate (SAR) for cell phones: What it means for you, Federal Communications Commission (FCC), //www.fcc.gov/consumers/guides/specific-absorption-rate-sar-cell-phones-what-it-means-you (last visited April 18, 2016).
37. SAR Information, //www.samsung.com/sar/sarMain.do (last visited April 18, 2016).
38. Corning Gorilla Glass, //www.corninggorillaglass.com/ (last visited April 15, 2016).
39. Explaining water and dust resistance ratings for your gadgets, //www.cnet.com/how-to/water-dust-resistance-ratings-in-gadgets-explained/ (2015).
40. Xperia™ Z5 Premium, //www.sonymobile.com/global-en/products/phones/xperia-z5-premium/ specifications/ (last visited April 15, 2016).
41. Discover the LG G5, //www.lg.com/us/mobile-phones/g5#G5Specs (last visited April 18, 2016).

42. Samsung is reportedly working on a phone with a crazy flexible screen that bends and folds, and it could launch in January, //www.businessinsider.com/samsung-flexible-foldable-phone-vidoes-2015-9 (2015).

43. Samsung's "Foldable Valley" could launch in January: Flexible phone uses a bendy plastic display to open and close like a book, //www.dailymail.co.uk/sciencetech/article-3236809/Samsung-s-Foldable-Valley-launch-January-Flexible-smartphone-uses-bendy-plastic-display-open-close-like-book.html (2015).

44. Nicole Arce, Samsung's New Graphene Technology will double life of your lithium–ion battery, //www.techtimes.com/articles/64353/20150629/samsungs-new-graphene-technology-will-double-life-of-your-lithium-ion-battery.htm (2015).

45. In Hyuk Son, Jong Hwan Park, Soonchul Kwon, Seongyong Park, Mark H. Rümmeli, Alicja Bachmatiuk, Hyun Jae Song, Junhwan Ku, Jang Wook Choi, Jae-man Choi, Seok-Gwang Doo, and Hyuk Chang, Silicon carbide-free graphene growth on silicon for lithium–ion battery with high volumetric energy density, //www.nature.com/ncomms/2015/150625/ncomms8393/full/ncomms8393 .html (2015).

Chapter 5

The Evolving Pay TV

Francisco Saez and Joaquín M. Lopez Muñoz

5.1 CUSTOMER NEED: ENTERTAINMENT

Starting from the mid-twentieth century, mass media and entertainment have become a major global industry with a market of nearly $2 trillion[1] that comprises production and distribution of movies, music, books, TV, radio, and, more recently, gaming (the latter, although a latecomer to the mix, has grown, for instance, to more than double the size of global box office revenues). In spite of the increase in competition for users' share of attention and the emergence of new ways to occupy leisure time on the Internet, TV watching remains hugely popular around the world (Americans sit in front of the TV set 5 h/day on average; in European Union the figure is 4 h[2]), and the associated market has a size of around $430 billion (60%[3] of which is Pay TV, the rest comes from advertising[4]) and growing. There are three factors that make *linear* TV (as opposed to other forms of video consumption) so appealing to people:

- *Linear TV is social.* Not only is TV watching an opportunity for families to meet and spend time together (as if gathering around a modern-day fire hearth), but also its potential for social connection extends beyond the home to friends, relatives, and workmates who are bonded by the experience of having enjoyed the same sports match or the last episode of a prime-time series. Being in the buzz is an important part of the sense of belonging in the community that every one of us aspires to. By contrast, picking up and watching a DVD (or a movie from an on-demand online service), even if a satisfying experience on its own, cannot be shared with the rest the day after.
- *Linear TV connects people to the world.* This is especially the case with live TV and news. Mass events (think of Olympics, the Oscar ceremony, the Super Bowl, the coverage of a major demonstration) are watched in real time or never. In underdeveloped countries, TV is frequently the only available window to the external world.
- *Linear TV is very well curated.* *Curation*, a term coming from the art exhibition world, refers to the process of selecting pieces of content and arranging

Digital Services in the 21st Century: A Strategic and Business Perspective, First Edition.
Antonio Sánchez and Belén Carro.
© 2017 by The Institute of Electrical and Electronics Engineers, Inc. Published 2017 by John Wiley & Sons, Inc.

them on display for maximum effect and enjoyment. In the world of TV, content is selected and scheduled by TV networks and broadcasters, who try their best, as their success depend on it, in programming their offer to maximize audiences in competition with other networks, especially at prime time. The fact that people spend hours on end watching TV is witness to the craft of the professionals in this industry—despite the oft-heard complaint that "there's nothing on TV."

Popular as it is, traditional TV in the twenty-first century is facing new challenges due to changes in user habits and the appearance of additional competition brought about by the emergence of Internet. Starting as a DVD rental company in the United States, Netflix decided in 2007 to move to online business and has become the global number one on-demand Internet video service with more than 80 million subscribers around the world as of 2016—and others, such as Amazon, are following suit. In the industry jargon, these new contenders are called by traditional companies, perhaps with a tint of disdain, *over-the-top* (OTT) players, as they leverage the Internet for distribution without having to set up any specific networking infrastructure on their own, which drastically reduces operational costs with respect to classical TV technologies (OTT players appear also as competitors to traditional companies in some other digital services discussed in this book). OTT services, and Internet in general, have risen the expectations of people about the immediacy and availability of content: Entertainment should be one click away on any screen, any time. In this context the restrictions posed by traditional TV, watchable only on the TV set according to a rigid programming guide, are increasingly alienating users, particularly among millennials. Unheard-of in the early 2010s, there is a growing (in some countries) segment of the population deciding to cancel their Pay TV subscription in favor of the more flexible (and frequently cheaper) video offer from OTT players, the so-called *cord-cutters*. Even though traditional TV and OTT services focus on different content offers, as we will see later, the threat of cord cutting has been enough of a warning for Pay TV operators to embrace the advantages provided by the Internet and IP connectivity and begin to extend the reach of their services with features such as multiscreen apps and *catch-up* access (live content stored on the cloud for on-demand consumption), which are now taken for granted in the United States and Europe and rapidly being adopted in the rest of the world. Moving from a fixed programming guide to a realm of thousands of on-demand content to watch calls for better and new curation capabilities, powered by automatic user profiling and recommendation technologies. Linear TV will not go away anytime soon, and will likely remain at the core of a more flexible offer, perhaps with increased focus on live events and prime-time windows, but the future of TV is definitely one of à-la-carte consumption, with many more choices and lower economic margins. The struggle for attention share in this new world will be won by those who can provide the best content to match each user's preferences, which is not only a matter of acquiring or producing the content itself but also requires knowing users down to the individual level dynamically over time.

5.2 CONTENT WARS

In essence, Pay TV can be defined as a service in which users *subscribe* with a distribution operator (called in the United States, for historical reasons, multiple-system operator or MSO) to a *package* consisting of a fixed selection of TV *channels* branded by a number of *TV networks* in charge of acquiring content from a *producer* (or producing their own) and curating it into a *programming* schedule. The resulting value chain is depicted in Figure 5.1.

In principle, Pay TV operators specialize in distribution and customer care, whereas content is the business of networks and producers. Typical types of contents are sports and live events, news, programs, movies, and series. The last two are the most interesting from the point of view of exploitation, as they are customarily enjoyed in a variety of ways not limited to Pay TV. So, movies from major studios go through a number of *exploitation windows* since they are released to the end of their useful lifetime, engineered to maximize the total revenues obtained by the producer.

- *Theatrical release.* At the beginning, the cost of the content for the user (which here is simply the money she pays to watch it at the theater) is maximum: From here on access is increasingly cheaper in accordance with the loss of luster as the movie ages.
- *DVD release.* After the movie leaves cinemas, it becomes available in DVD format. This window can be subdivided into DVD *purchase* and *rental*; sometimes the former (which derives more benefit for the studio) is a little ahead in time. DVD and its predecessor technologies like VHS were incredibly profitable for studios from the 1980s, but the emergence of online streaming has caused this consumption format to plummet, even if replaced by more advanced physical technologies like Blu-ray Disc. With some reluctance, studios are giving way to lower margin online consumption.
- *EST (Electronic Sell-Through).* This is the online equivalent of DVD purchase and has been slowly promoted by studios, albeit with some reluctance due to security and margin concerns, as DVDs go out of vogue.
- *TVoD (Transactional Video on Demand).* It is the online equivalent of DVD rental. EST and TVoD share approximately the same window as physical DVD marketing.

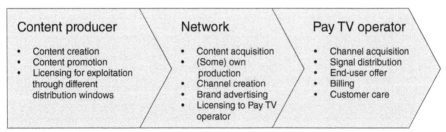

Figure 5.1   Pay TV value chain.

- *Pay TV.* Around a year after the movie was first released in theaters and has been sufficiently exploited in purchase/rental mode, TV networks are granted rights to acquisition and exhibition through Pay TV only (as the cost of content at this point in time makes advertising-supported distribution in free TV impractical). In some cases the network is allowed (or can pay an extra fee) to store the movie on the cloud for on-demand access in a short time frame following its airing on TV: This is known as catch-up consumption. Pay TV operators are increasingly pushing for the rights to store catch-up content themselves, and there are currently no industry-wide rules about rights in this hot arena.

- *SVoD (Subscription Video on Demand).* This window opens up at around 2 years after theatrical release and generally precedes that of Free to Air by a short time. SVoD content is made part of large movie catalogs exploited on a per-subscription basis by OTT players such as Netflix. SVoD plays the trade-off between content freshness (poor) and breadth (wide).

- *AVoD.* An alternative to SVoD is ad-supported VoD, where on-demand content is available free of charge with some advertising interspersed along the streaming session. Broadcaster-owned service Hulu is the prime example of this type of window. AVoD roughly coincides in time SVoD. Do not confuse AVoD with ad-supported video streaming services such as YouTube, which typically exploit user-generated clips, music videos, and types of content other than premium movies.

- *Free to air.* At the end of its useful lifetime, the price of the content has decreased to a point where nonpremium networks and broadcasters can acquire it for distribution over free TV, whose revenues depend solely on advertising (Figure 5.2).

The exploitation windows of series are not exactly the same as those of movies (there is no theatrical release, for instance), but the gist of it remains the same: purchase/rental and broadcasting through Pay TV precede availability in subscription-based online services and free TV. This would seem to preclude fiercer competition

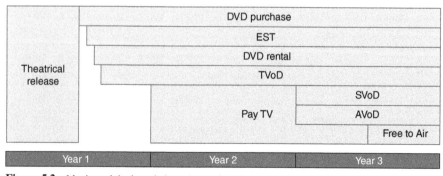

Figure 5.2   Movie exploitation windows (approximate).

between Pay TV and OTT services, but the rapid growth of the latter and the relative inexpensiveness of reaching the user through Internet have set everyone in pursuit of a larger share of the Pay TV and video markets, in a series of movements that we can humorously dub the Content Wars:

- OTT players, particularly those with greater financial muscle, are trying to move out of the 2-year-old SVoD window by bidding for exclusivity of content to compete with networks at the Pay TV window and by producing their own content for direct online consumption. Leveraging to the maximum the on-demand nature of their service, Netflix introduced in 2013 the possibility of *binge-watching* complete seasons of their home-produced series, that is, making all the episodes available at once for users to enjoy one after another in one swoop, which has proved popular among a large segment of the audience.

- Some TV networks, for example, HBO,[5] are beginning to offer their channels directly to users, which can be more profitable than going through Pay TV operators as intermediary (though this has to be balanced with the costs of customer care and infrastructure for Internet distribution). It allows anyone to pick up the network offer without having to subscribe to a larger Pay TV package with filler content she is not interest in. This goes in line with the general trend to offer more flexibility to users when deciding their content mix, and is naturally most appealing to premium networks who feel can compete on their own brand.

- Sports producers, such as the MLB[6] or the NBA, are also challenging the *status quo* and offer their real-time content online, usually integrated into very interactive applications with additional stats, multiple camera views, and so on targeted at hardcore fans. Again, this is meant to capture users not willing or unable to pay for an expensive TV package, and those in countries where no network has acquired the distribution rights because this particular sport is not popular enough.

- Some Pay TV operators seek to secure their position and establish a greater differentiation with competing operators by producing their own TV channels and content. Two examples out of many are Sky UK and Movistar+ Spain, each the dominant Pay TV operator in their country.

We do not dare predict the outcome of all these operations around content production and exploitation, but they will certainly result in a more complex scenario for users to choose from, which might eventually open up the opportunity for OTT to offer integrators replicating the role played by Pay TV operators today—device manufacturers such as Apple (Apple TV) and Amazon (Fire TV) would be certainly happy to step in, but the list of candidates includes other Internet players such as Google as well as Pay TV operators themselves. Also, it must be noted that content production is a very expensive activity, which naturally favors the biggest in place: As Content Wars are fought and the video value chain becomes more and more blurry, only a very few might survive at the end, which would stymie further competition.

5.3 AGGREGATION VERSUS DIVERSITY

People have the two contradictory desires to simplify their lives and to have many options to choose from. In a series of now famous experiments, psychologist Barry Schwartz[7] purportedly proved that people tend to choose suboptimally and be more anxious as the number of options they are presented grow (an effect he calls "the Paradox of Choice"). This is in opposition to the generally celebrated freedom that Internet provides when it comes to entertainment selection. Which line of thought is right?

A major theme in the 2010s is the strong trend toward service bundling, which is perceived as a win–win scenario by both the user, who has centralized customer care, fewer bills to deal with, and is frequently offered a price discount on the bundle, and the service provider, as bundling has proved to result in higher customer retention and lifetime value. Telco operators are increasingly offering *triple play* (3P) bundles where Pay TV is added to the basic voice and data services, and conversely Pay TV operators look for ways of providing connectivity services either by leveraging their cable infrastructure or directly merging with telco companies. Experience in the United States, where Pay TV is a saturated market with more than 80% penetration, shows that only 3P operators are able to grow in a scenario where cord cutting is slowly eroding the market for non-3P traditional MSOs.

Subscription-based services, examples of which are Pay TV and SVoD, the latter first introduced by Netflix, have a "free lunch" psychological effect on users (also known as sunk-cost effect), who do not have to go with the value–cost assessment process associated to each consumption in a transactional-mode service. This can even result, when the first phase of compulsive usage subsides, in a subscription service having more return than the equivalent transactional service would obtain for the same consumption profile—which is ultimately dangerous if users come to perceive that subscription is not getting them enough value for their money, particularly so in the online world where opting out is extremely easy. With the increase of online video offerings, Pay TV might be approaching this tipping point signaled by cord cutting or, in less extreme cases, *cord shaving* (downgrading to cheaper, core TV packages while going online to complement the entertainment offer). Analysts have pointed out to the existence of a still small segment of the population they call *cord-nevers*,[8] youths and young adults who never subscribed to Pay TV and do not plan to do it since their entertainment needs are sufficiently covered online. It is not clear yet, however, whether cord-nevers will continue to stay away from Pay TV for the rest of their lives or else will progress to the more conservative consumption habits of their parents once they themselves settle down or raise a family, and so on.

As for diversity in video offers, OTT players have made it a sales pitch point to advertise the *size* of their catalogs (which in itself is not such a good indicator of catalog *quality*). In practice, though, consumption in these services follow a Pareto-like distribution where most activity concentrates in just a handful of very popular titles, but even so users find the availability of massive amounts of long-tail content extremely appealing. Navigating huge movie catalogs calls for personalized curation capabilities that can prevent content from simply not being discovered ever.

On closer analysis, the two aspirations for simplification and freedom of choice are not that contradictory after all: users want to simplify *service management* while enjoying *consumption diversity* matched to their personal (and their families') preferences. With these considerations in mind, we can venture to signal some key characteristics of the successful Pay TV of the future:

- Integrated with telco services in a 3P bundle.
- Oriented toward subscription to linear TV (and maybe also SVoD) packages organized in a very granular offer easy to opt in and out.
- Offering flexible consumption with catch-up assets (broadcasted content that can be enjoyed at a later time on an on-demand basis).
- Universally available through multiscreen clients on any user device.
- Heavily curated by the Pay TV operator (not merely based on network's own curation) aided by user profiling and personalized recommendation technologies.

5.4 THE ROLE OF ADVERTISING

The relationship of customers with TV advertising is one of love and hate. On one hand, advertising revenues contribute to lower Pay TV bills and can even wholly support free TV, which is prevalent in some countries. On the other hand, ads (unless particularly relevant or beautifully crafted) interfere with and degrade the TV watching experience. European countries are typically less tolerant of advertising than those of America and the rest of the world.[9] The value chain of the $180 billion TV advertising market is a complex one, involving many agents from *advertisers*, *media agencies* and *media buying companies*, TV networks, and, in some cases, Pay TV operators as well. Without entering into further details, suffice it to know that the value of the chain is determined by the availability and ownership of so-called *inventory*, that is, the TV slots where commercials from advertisers can be inserted for broadcasting, whose price greatly depends on audiences as determined by *audience measurement* companies. In general, inventory is owned by TV networks, who consequently play the dominant role in the advertising value chain and do not let Pay TV operators partake of the business, although in the United States MSOs are granted by law a small part of network inventory (around 4 min per hour) to exploit on their own, in many cases for local ads but also in federation with other MSOs for larger reach (this curious arrangement is far less frequent in the rest of the world). Another traditional way for a Pay TV operator to enter the advertising chain is to produce its own channels, which is in general only feasible for the largest companies.

Much can be done in order to advance the relatively old schemes of TV advertising for greater effectiveness and profitability in a world of interactivity and online services. Audience measurement can be made more precise and detailed by taking advantage of connected TV devices (set-top boxes, STBs) used by Pay TV operators. In particular, Pay TV operators now have the means of supporting networks in the implementation of *addressable advertising,*[10] by which general-audience commercials are transparently replaced by more relevant ones specifically targeted at predetermined geographical areas or population segments; replacement can even take place for catch-up content, whose inventory is basically available as advertisers typically buy just the airing time, and thus cannot claim ownership of their presence in recorded content. Addressable advertising increases the value of TV inventory and disrupts the business value chain since Pay TV operators become an integral part of it as providers of user intelligence. Besides linear TV advertising, on-demand content, be it TVoD, SVoD, or again catch-up, can be used for ad insertion in the form of *pre-rolls* (before watching begins) or, more intrusively, *mid-rolls*. In general, connectivity multiplies the possibilities for securing or augmenting inventory as part of the user interface itself, in the *second screen* simultaneously with advertising on the TV set, via interactive proposals where users can be called to action from a commercial to buy online, and so on. The limitations are not so much technical as political, since any new possibility in the realm of TV advertising involves significant changes to the current status quo controlled by networks. Another problem is related to pricing schemes: online advertising, which has grown roughly to the same size as TV advertising, plays by different rules for ad measurement and inventory selling, and no one has yet found a convincing way to reconcile both worlds. There is little doubt in the industry that these limitations will be eventually overcome: The TV advertising landscape 10 years from now will not look anything like the present.

5.5 TECHNOLOGY: SATELLITE, CABLE, AND IPTV

Pay TV started with physical links into coaxial cable network managed by Pay TV cable operators delivering analog channels through its controlled footprint only for subscribed customers.

The two main drivers for the early evolution of Pay TV operator services have been fighting TV signal piracy and enriching the user experience around linear channel packages that are basically shared among all competitors and thus offer little room for differentiation.

All the technology enablers to make Pay TV possible could be named "Pay TV or Video Platforms" in the sense that if we name "platform" in the technology industry, almost whatever set of systems we do not know how to call them are lumped all together as a platform (Figures 5.3 and 5.4).

5.6 PAY TV TECHNICALL KEY COMPONENTS

Devices. Pay TV is no longer an isolated service in a TV set but an entertainment experience across a set of devices, from consumer electronics owned by the customers such as smartphones, tablets, video game consoles, or Smart TV to devices fully managed by the operator, the most common of which is the set-top box. Such

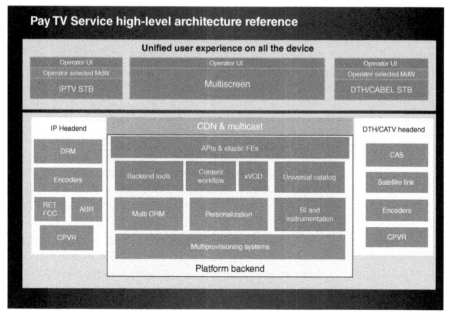

Figure 5.3 Pay TV platform high-level architecture reference.

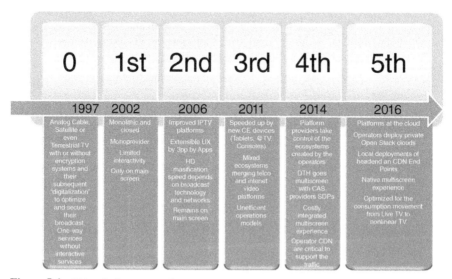

Figure 5.4 Video platforms generations summary.

experience has to be seamless across the different screens leveraging on the Pay TV platform backend capabilities.

Headends are the set of components responsible for the reception of linear TV contribution channels and their securitization against piracy for distribution through a defined technology. The most common distribution technologies are Direct-to-Home satellite (DTH), Cable TV through HFC networks (CATV), IPTV through operator-owned ultrabroadband networks, and over-the-top Linear TV for unicast streaming typically delivered through Content Delivery Networks (CDN). The most relevant components of any headth are the following:

- *Encoders*, which compress video and audio at the selected quality under a defined standard coding technology supported by the devices for its decoding. Adaptive bitrates packagers (ABR) and other enablers for IP multicast reliability are also quite common (FCC fast channel change and RET error correction systems).
- *Content protection systems* compatible with the devices. These stand[11] for the collection of technologies that technically enable the definition and enforcement of secure content transportation as well as secure content licensing. For IP networks they are Digital Rights Management systems (DRM) or Conditional Access Systems (CAS) for pure one-way broadcast services (DTH/CATV CAS).
- Other service enablers such as **Cloud PVR** to manage customer content recordings on the network instead of Personal Video Recorder-enabled STBs.

Platform backend core refers to those technological components necessary to run the logic of the TV service. The most significant service enablers are the following:

- *Provisioning systems:* Responsible for the customer management, usually connected to higher level operator business support systems (BSS).
- *xVod enablers:* Content workflow including VoD assets encoding, asset management systems to manage content catalog.
- *MultiDRM[12]:* Entitlement system to guarantee maximum compatibility with devices.
- *Recommendation engine* for a personalized curated contents catalog browsing experience.
- *Business intelligence (BI):* Component, responsible for big data management, provides insights in order to generate collaborative filtering recommendations, creating cluster of users according to their consumption preferences.

5.7 EVOLUTION OF INTERACTIVE PAY TV TECHNOLOGIES

Pay TV platforms, which started mainly concerned with distribution and the associated broadcast technologies (cable, DTH, IPTV, and more recently online or OTT video), have progressively included additional components to enrich the Pay TV and make it more interactive.

Interactive TV means an evolution on passive TV channel from just zapping by to enabling access to further services, for example, VoD contents, customized program guide with enriched metadata, content search, and any sort of side applications available on TV.

Along the course of evolution of interactive Pay TV we can identify a number of significant advances that we can group into generations for the sake of better understanding.

5.7.1 Generation Zero: Early 1950s to Late 1990s

We can consider as generation zero those pay TV services based on analog cable, satellite, or even terrestrial TV with or without encryption systems and their subsequent "digitalization" to optimize and secure their broadcast. Their lack of interactive services due to their intrinsically one-way nature left them only as a reference but not the missing link of the evolution of Pay TV.

The almost "universal" reach of DTH services due to the broad coverage of DTH satellites and their ease of installation has been, and is, the key to success of this technology, now enhanced by hybrid STBs, which combine satellite signal and OTT video services in a single device.

5.7.2 First-Generation Video Platforms: Early 2000s

The interactive Pay TV started leveraging on the IP broadband with IPTV. The IPTV service was the reaction of the telco operators to the cable companies that started to offer voice and broadband through their cable TV networks, challenging the telco's voice and broadband business. Cable operators or telco companies are no longer significantly different, as both are now customarily offering triple play services. Let us focus on the invention of the IPTV.

IPTV[13] was defined by the Alliance for Telecommunications Industry Solutions (ATIS) as follows: . . . *the secure and reliable delivery to subscribers of entertainment video and related services. These services may include, for example, Live TV, Video On Demand (VOD) and Interactive TV (iTV). These services are delivered across an access agnostic, packet switched network that employs the IP protocol to transport the audio, video and control signals. In contrast to video over the public Internet, with IPTV deployments, network security and performance are tightly managed to ensure a superior entertainment experience, resulting in a compelling business environment for content providers, advertisers and customers alike.*

The IPTV world started with monolithic and almost single provider platforms, fully oriented to provide subscription linear channels and Pay per View events leveraging on IP multicast delivery of audio and video using RTP/RTCP over managed networks, 100% under the operator control and with guaranteed QoS.

The service can only be enjoyed on the main TV set at home, which is equipped with an IP set-top box directly connected to the broadband modem router.

The first interactive service to be included in Pay TV was VoD. Without real interactivity at hand, DTH and cable operators had managed to provide a limited sort of VoD experience by broadcasting blockbuster movies through dedicated Pay per View channels: Users could purchase and watch the content at a number of programmed windows, but without any possibility to manage the video playback by pause, rewind, or fastforward. Push VoD to PVR STBs was another breakthrough but with very limited success due to technical limitations. First versions of real IPTV VoD, with limited catalog availability, were provided by means of unicast RTSP. Scalability of the VoD service necessitates the distribution of points of presence (PoPs) of the IPTV service very close to the customers, resulting in a content distribution system with limited capacity for video management.

5.7.3 Second-Generation Video Platforms: 2004–2010

IPTV services evolved quicker than DTH and cable, including differential features for enhanced TV experience.

- Instant/fast channel change.
- Multiroom PVR, which leverages the *local area network* (LAN) at home to connect all STBs in the household to a single main PVR.
- Picture in picture. One channel is displayed on the full TV screen at the same time as one or more other channels are displayed in inserted windows, even as a channel mosaic.
- STB software stacks become a little more open and allow the service to be extended by means of interactive Apps development frameworks.

HD TVs and the evolution of the quality of the live TV brought by the HD channels are a key driver in the evolution of the pay TV services, too. DTH, CATV, and IPTV started to deliver HD contents.

- IPTV faced a serious limitation for the universalization of HD channels on the access network due to lack of bandwidth, which limited HD Pay TV to ultrabroadband networks (VDSL and FTTH at 20+ Mbits downstream). On the other hand, where ultrabroadband was present there were no practical limitations to the number of HD channels that could be delivered.
- The initial investment for DTH and cable operators to introduce HD was higher due to the lack of support of MPEG4 on deployed STBs and the high cost of the satellite transponders. Availability of HD was in principle universal, provided the user upgraded to a newer STB, but typically the number of HD channels was lower than in IPTV.

5.7.4   Third-Generation Video Platforms: 2011–2013

New connected devices other than PCs (smartphones, tablets, consoles) began to appear at home, providing people with additional venues for video watching, even while sitting in front of the TV set.

Content watchable on those devices was provided by "Internet content providers" such as YouTube, Netflix, and catch-up sites for most popular TV channels such as Hulu. Video OTT was born. No general standard for OTT video distribution existed, and each manufacturer proposed their own set of technologies: for example, Apple HLS, MS Smooth-Streaming, PlayReady, Widevine, and Adobe HDS.

Most of Pay TV video platform providers of the time were not ready to react to this phenomenon in order to provide integrated platform solutions to the operators. The IPTV operators found "quick and dirty" solutions to circumvent this by using platform-as-a-service solutions managed by their Internet services providers, completely isolated from their existing Pay TV services.

The management costs grew due to the inefficiencies to run, integrate, and evolve these service platforms in parallel to the main screen one. The development of the Pay TV "video ecosystems" started more or less managed by the Pay TV operators subsidized by main screen revenues versus the uncertain return of investment of the second screen, and mainly to protect their customer base against the threat of the pure OTT operators.

5.7.5   Fourth-Generation Video Platforms: 2013

Pay TV technology platform providers started their late reaction to OTT proliferation in different ways trying to avoid the risks for their revenues of these heterogeneous "video ecosystems" and under the pressure of Pay TV operators, who needed to reduce the operational costs of integrating external OTT services themselves.

Video technology providers evolve their product portfolio to include the missing OTT services. The development pace of the platforms for Telco is not enough to meet the expected time to market the operators were requesting.

New players coming from the OTT world started to promote end-to-end multiscreen IPTV solutions. In the DTH and cable arena, conditional access suppliers[14] jumped into the IP world providing Service Delivery Platforms (SDP) with proprietary DRM solutions to offer multiscreen services. Hybrid services started to be a reality as a convergence path of DTH and IP Interactive services.

Some sort of integrated user experience of interactive services across the devices started to be possible and requested by the customers. From the main STBs to any other connected device at the home, video OTT enables the "cross-screen" entertainment. New devices and new use cases[15] are now possible with home media gateways with transcoding capabilities and multiple built-in tuners. Pay TV operators started to deploy their own operator CDNs as the most effective way to guarantee video quality in a more efficient way than former IPTV video pumps.

5.7.6   Fifth-Generation Video Platforms: 2016–Current

The fourth-generation video platforms, with built-in multiscreen features, are already relying to a greater or lesser extent on virtualization technologies (VMware, HyperV, etc.). The next leap in the evolution of video is the introduction of *cloud computing* and the possibilities that elastic computing provides to video services. Video services can run on public clouds such as Amazon AWS or MS Azure, as the pure OTT video services providers are doing since the very beginning (Netflix, Amazon Instant Video).

Video platforms evolve their software architectures to be deployed on private Operator Clouds based on *OpenStack*[16] in order to guarantee high availability and business continuity based on georedundancy and data center cloud distribution. The DevOps model is introduced, which shortens time to market, automates QA testing, and allows for controlled deployment of new features (*A/B testing*).

Pay TV services leverage on big data technologies to understand and predict the behavior of customers, which can be further used by built-in or third party content recommendation engines or targeted advertising services.

TV headends and other enablers to ingest live channels into the network will continue to be locally deployed, while the rest of the services are managed from the cloud. Platforms evolve by optimizing video delivery in combination with the CDNs in order to support the increase of the unicast video consumption due to VoD. Unicast live OTT traffic *offloading*, which leverages multicast ABR[17] solutions, is a key differential factor for telco operators with respect to pure OTT competitors.

5.8   VIDEO DEFINITION

For over a century, we have been able to capture moving images with increasingly better quality. To achieve the level of acuity that the human eye is able to perceive, further improvements in visual technology are still required.

Visual quality of moving images can be broken down into four parts[18]: resolution, frame rate, color space, and dynamic range (intensity difference between bright and dark).

Video resolution is a key component of user experience. High Definition has been mainstream for the last few years and the next step is Ultrahigh Definition (UHD 4 K, 8 K), but focusing on resolution alone is not enough, and will fail as we saw 3D TV failed. Under usual viewing conditions, the human eye simply cannot perceive the difference in terms of resolution improvements beyond HD. There is room for improvement in Ultra-HD video, however, but it comes from other aspects of perception: high dynamic range (HDR), wide color gamut (WCG), and high frame rate (HFR).

High dynamic range boils down to having deeper blacks and brighter whites. Adopting HDR implies upgrading the entire technological chain from the camera to the display and all the way in between, in production sites, in delivery formats, and distribution to the consumer.

Increasing quality has a price in terms of generating more raw data, which translates to distributing more information, so one of the challenges is to compress it down to manageable limits when video is transferred over a satellite link or a broadband connection. This requires making some compromises on the video quality, although algorithmic improvements in codec evolution (from H.264 to HEVC or VP9) and careful tuning of compression parameters can make degradations in quality largely acceptable or even hardly noticeable by viewers.

Massive adoption of HDR requires next-generation UHD TV displays with OLED or Quantum Dots technologies. Associated power consumption concerns are also being addressed by TV manufacturers. Isolatedly or in combination with improvements in frame rate (120 fps) and wider color gamut, it will be possible to reproduce "the real world" better than ever.

There are several HDR technologies competing to become the *de facto* standard. Although all of them are similar in quality and keep getting better, the industry needs to reach some sort of convergence in order to have a single and future-proof UHD TV. Blu-ray Disc Association, ARIB, Technicolor, and Dolby Vision are the most active on this front. The way each technology manages backward compatibility with current infrastructure and Standard Dynamic Range (SDR) displays is probably the hottest topic. *Ultra-HD Forum* can facilitate and coordinate standardization activities across the plethora of standards defining organizations involved throughout the chain.

HDR UHD is expected to be more true to life than anything you have ever seen on a TV, giving viewers such a realistic and stimulating TV viewing experience that you might even want to reach for your sunglasses! But please do not hesitate to get out of your living room to get some fresh air!

Going beyond UHD, Virtual Reality (VR) artificially creates sensory experiences, which can include sight, sound, and touch. VR offers computer-generated images that appear on head-mounted displays (HMD). Oculus Rift, Samsung Gear, HTC Vive, and Google Cardboard are the most popular.

Besides the almost unlimited possibility that VR will provide to create new interaction ways across many industries by augmenting the reality, focusing on Pay TV, immersive contents, and 360 degrees user interfaces are clear breakthroughs. New content categories are showing up closer to the video games than to the movies. It will be challenging for Pay TV operators to compete with OTT players that are promoting 360° user-generated content services and are less reluctant to adopt new technologies.

ACRONYMS

3P   triple play

ABR   adaptive bitrates packagers

ATIS   Alliance for Telecommunications Industry Solutions

AVoD   Ad-supported Video on Demand

BI   business intelligence

BSS   business support systems

CAS   conditional access systems

CDN   Content Delivery Networks

DRM   digital rights management

DTH   Direct-to-Home satellite

HD   High Definition

HDR   high dynamic range

HFR   high frame rate

HND   head-mounted displays

MSO   multiple-system operator

OTT   over-the-top

PoPs   points of presence

PVR   personal video recorder

SDP   service delivery platforms

STBs   set-top boxes

SVoD   Subscription Video on Demand

TV   television

TVoD   transactional video on demand

UHD   Ultrahigh Definition

VR   virtual reality

WCG   wide color gamut

NOTES

1. McKinsey & Company (2015), Global Media Report 2015, //www.mckinsey.com/~/media/mckinsey/dotcom/client_service/media%20and%20entertainment/pdfs/mckinsey%20global%20report%202015_uk_october_2015.ashx (retrieved June 2016).
2. IHS (2015), Brits watch less tv than ever before; Italians watch the most, IHS Says, http://press.ihs.com/press-release/technology/brits-watchless-tv-ever-italians-watch-most-ihs-says-infographic (retrieved June 2016).
3. Broadband TV News (2016), Global pay TV subscribers to surpass 1.1 billion in 2020, //www.broadbandtvnews.com/2015/03/26/global-pay-tv-subscribers-to-surpass-1-1-billion-in-2020/ (retrieved June 2016).
4. Research and Markets (2015), Global TV advertising, //www.researchandmarkets.com/research/4tqhvx/global_tv (retrieved June 2016).
5. Baumgartner, J. (2015), "HBO now" goes live, Multichannel News //www.multichannel.com/news/tv-apps/hbo-now-goes-live/389517#sthash.gEGufCmc.dpuf (retrieved June 2016).
6. MLB.com (2012), Ten years ago today, MLB.TV debuted, http://m.mlb.com/news/article/37372302/ten-years-ago-today-mlbtv-debuted (retrieved June 2016).
7. Schwartz, B. (2004), *The Paradox of Choice*, Harper Perennial.

8. Advertising Age (2011), "Cord-avoiders" to reduce cable and satellite TV rolls in 2012, analysts say, http://adage.com/article/mediaworks/cable-satellite-tv-subscriber-rolls-shrink-2012/231216/ (retrieved June 2016).

9. Nielsen (2009), Nielsen Global Online Consumer Survey, Trust, Value and Engagement in Advertising, //www.nielsen.com/content/dam/corporate/us/en/newswire/uploads/2009/07/trustinadvertising0709.pdf (retrieved June 2016).

10. AdvertisingAge (2014), The CMO's guide to addressable TV advertising, http://adage.com/article/cmo-strategy/cmo-s-guide-addressable-tv-advertising/291728/ (retrieved June 2016).

11. ATIS Definition of DRM, //www.atis.org/glossary/definition.aspx?id=9058 (June 2016).

12. Viaccess Orca (2015), White Paper: Adopting DASH & Multi-DRM for Video Delivery—The Great Opportunity for Change, http://blog.viaccess-orca.com/industry/white-paper-adopting-dash-multi-drm-for-video-delivery-the-great-opportunity-for-change/ (last accessed June 2016).

13. Alliance for Telecommunications Industry Solutions (ATIS) IPTV Exploratory Group on 2005.

14. Nagra (2014), Nagra enables over one million active devices for Canal+ Spain's over-the-top service, //www.nagra.com/media-center/press-releases/nagra-enables-over-one-million-active-devices-canal-spain%E2%80%99s-over-top (June 2016).

15. DirecTV Genie TV, //www.directv.com/technology/genie (2016)

16. Openstack, Open source software for creating private and public clouds, //www.openstack.org/ (June 2016).

17. CableLabs (2014), IP Multicast Adaptive Bit Rate Architecture Technical Report, //www.cablelabs.com/wp-content/uploads/specdocs/OC-TR-IP-MULTI-ARCH-V01-1411121.pdf (June 2016).

18. Ericsson (2015), Understanding Ultrahigh definition television, //www.ericsson.com/ar/res/docs/whitepapers/wp-uhd.pdf (retrieved June 2016).

Chapter 6

Enterprise: From Machine-to-Machine Connectivity Toward Internet of Things

Machine-to-machine (M2M) is a form of mobile data connectivity devoted to connecting machines, that is, autonomous devices without human intervention. Two prominent examples of vertical markets are automotive (cars) and utilities (smart meters). In 2014, it represented approximately 3% of world mobile connections, about a quarter of a billion lines.[1] As opposed to traditional mobile broadband, typically M2M provides a low average revenue per connection, with more limited data consumption, while the contracts are for longer periods, lower churn, and might involve very large number of connections, several millions, for example, a 53 million smart meters contract in the United Kingdom. Beyond basic connectivity, in which the customer pays only for data consumption, it is useful to have managed connectivity, in which a management portal allows the service provider to control all the lines (activations, consumption per line, etc.), which is an important function when the number is large. Another common feature of M2M deployments is that they are often done for multinational companies, but there are no global telecommunications operators with worldwide coverage. Beyond traditional roaming, it is common that several operators (sometimes as part of established alliances) provide a joint service. Embedded SIM (Subscriber Identity Module) specifications allow over the air provisioning and change of operator subscriptions. In other words, the SIM card is included as part of manufacturing process and operator selection is done remotely.

M2M has different traffic patterns, which require changes in the underlying network. An example is Low-Power-Wide-Area network, which is a network optimized for M2M. They are designed to minimize the power consumption so that M2M devices (which many times cannot be plugged to electricity) can run on a battery for a very long time, for example, 10 years from a single AA cell.

Digital Services in the 21st Century: A Strategic and Business Perspective, First Edition.
Antonio Sánchez and Belén Carro.

Another feature is to provide long-range connectivity for very remote areas, thus minimizing the deployment costs for universal coverage. There are different competing technologies in this space, and cellular networks are defining new standards (as part of 3GPP latest release 12 and upcoming 13) to address this new area of applications.

Beyond connectivity, M2M enables value-added applications in different vertical markets. Traditional examples of applications come from automotive sectors such as fleet management and connected car, for example, 7 million subscribers[2] of General Motors's OnStar service, which pioneered the concept. But every industry can benefit, and a new term coined Internet of Things is used for connecting every object and providing applications on top of this connectivity.

6.1 CUSTOMER NEED: REMOTE AUTOMATION

In the world there are billions of devices that could become smarter if a connectivity layer is added to their capabilities. In the same way that Internet revolutionized computers (and later mobile devices) used by people, data connectivity for machines can transform significantly enterprises. Companies need to improve their productivity, and enabling remote communications for their machines can reduce their costs and also offer better services to their customers. This applies to many different industries but we will focus the discussion on two prominent examples: utilities and automotive.

6.1.1 Utilities

Utilities bill their customers according to meters installed in their premises. Traditional ways of reading meters require personnel to travel periodically (e.g., every 1/2/3 months) to the customer premises and manually read the meter by visual inspection and register it in the charging system. With M2M, mechanical meters are replaced by electronic meters that include a connectivity module that sends the data directly to the central server. This remote reading saves significant costs, reduces errors, allows instantaneous readings, and prevents theft and tamper. This applies to water and gas, but in the case of electricity, the productivity gains are even greater. Electricity costs are variables as a function of demand. In peak times, less efficient generation sources have to be used and this means much higher production costs. With smart connected meters, utilities can charge the customer according to the cost (e.g., different tariffs per hour and day—weekday versus weekend). This variable price can in turn flatten the demand, with customers trying to consume less on peak hours and more on cheaper ones. Although in some cases electricity companies are using their own networks and technologies to connect smart meters, power line communications (e.g., Iberdrola in Spain), radio technologies, are well positioned to serve the market. As a matter of fact, in 2013, the largest M2M contract ever signed was for smart meters in Great Britain,[3] awarded by the government (Department for Energy and Climate Change) under the Smart

Metering Implementation Programme. It involves 53 million electricity and gas meters by 2021. The United Kingdom operator O2 (at the time part of Telefonica, which in 2015 reached an agreement to be sold to Hutchinson Whampoa) won two (Central and Southern) of the communication service provider regions (out of 3) with a contract worth 1.5 billion pounds over 15 years—mostly based on cellular but with alternative technologies for zones with bad coverage.

6.1.2 Automotive

Automotive is a huge industry, surpassing 1 billion vehicles in operation worldwide already in 2011.[4] Cars are expensive machines and it is usually quoted as the third place where people spend a lot of their time (after home and office). Again, connecting them can optimize the automotive industry and the lives of the motorists. A typical use case for cost savings is to remotely deliver vehicle diagnosis information to a workshop in case of failure, or even better before it happens to preventing major damages. From a safety perspective, an emergency call by the car itself can save lives in case of accident, alerting emergency center directly with important information about location and even about the nature of the accident itself (e.g., car speed, deceleration) that could be useful for medical purposes. A more general productivity gain can be achieved with connectivity, since it allows manufacturers to obtain very useful data about the vehicle status and its operation, thanks to analytics that can improve the overall manufacturing process. As of today, it could be said that every major car manufacturer has reached an agreement with a cellular operator to progressively include connectivity in the new cars sold, at least in the high end. As an example, AT&T registered 684,000 connected cars in first quarter of 2015.[5] AT&T's connected car platform for automakers is called *AT&T Drive* and includes apps such as remote diagnostics.

6.2 BASIC CONNECTIVITY AND MANAGED CONNECTIVITY

In its basic form, M2M just provides connectivity. The machine is connected, either through an embedded or an add-on module that includes the radio chipset. In the case of cellular, the module includes a smartcard that authenticates to the network, and is provided by the operator. The customer consumes the data and is charged accordingly. In these Internet times, data are typically transmitted through IP networks to the backend systems of the company that owns the machine.

6.2.1 Operators

A growing number of telecom operators are disclosing in their quarterly financial results the number of M2M lines they serve and sometimes even the revenues, as shown in Table 6.1 (corresponding to first quarter of 2015). As can be seen, the

Table 6.1 Machine-to-Machine Key Performance Indicators (Units, Revenues) for Telecom Operators

Telecom operator	Units	Growth	Revenues	Growth	Region
China Mobile[a] AT&T[c]	43 million M2M (end 2014) 22 million connected devices (including session-based tablets)	36% yoy 19% yoy			China The United States
Vodafone	21 million M2M	34% yoy		25%	Parts of Europe, Africa, Middle East, and Asia Pacific
Verizon[c]	>15 million IoT (2014)	40–204% (manufacturing: 204% Finance and insurance: 128% Media and entertainment: 120% Home monitoring: 89% Retail and hospitality: 88% Transportation and distribution: 83% Energy and utilities: 49% Public sector/smart cities: 46% Health care and pharmaceutical: 40%)		45%	
Telefonica[d]	10 million M2M	28% yoy	€ 37 million (quarter)	25%	Latin America, Spain, Germany, and the United Kingdom
Deutsche Telekom[e]	4 million M2M (business prepay customers) 5 million M2M (business prepay customers)	34% yoy 19% yoy			Germany The United States

Telstra[f]	1 million M2M (end 2014)	27% yoy	AU$55 million (half year)	17%	Australia
Orange	3 million M2M[g]	35% yoy			France

[a] 2014 Annual Results Presentation (Investor Relations. Presentations/Webcasts). China Mobile. //www.chinamobileltd.com/en/ir/webcasts.php (March 19, 2015).

[b] Quarterly Earnings—1Q 2015. AT&T, //www.att.com/gen/investor-relations?pid=268 (April 22, 2015).

[c] Internet of Things gains momentum among businesses, New Verizon Report reveals. Verizon, http://news.verizonenterprise.com/2015/02/verizon-internet-things-report-2015/ (February 23, 2015).

[d] Quarterly Results 2015. Telefonica, //www.telefonica.com/en/shareholders-investors/html/financial_reports/resultados2015.shtml (May 14, 2015).

[e] Financial results for first quarter 2015. (*Note:* M2M figures for Germany and the United States, other countries not disclosed.) Deutsche Telekom, https://www.telekom.com/15Q1 (May 13, 2015).

[f] Investors. Latest results. Half year Results 2015 (period ended December 31, 2014). Telstra, //www.telstra.com.au/aboutus/investors/latest-results/ (February 12, 2015).

[g] Orange investors data book, KPIs Q1 2015. (*Note:* M2M figures for France, other countries not disclosed.) Orange, //www.orange.com/en/content/download/30283/842657/version/1/file/Book+KPIs+Q1+2015+VD.pdf (April 28, 2015).

growth is healthy, for example, about 30% year-over-year (yoy). All figures are rounded to the closest integer (lower in case of half point). It has to be noted that the definition of the accounting term might differ among operators—that is why the used term by each is quoted.

These players represent approximately 40% of the total market in terms of connections. Other major mobile operators that apparently do not disclose data regularly include America Movil, NTT Docomo, Softbank, KDDI, Singtel, China Telecom, Bell Canada (BCE), and China Unicom (top 18 mobile service telecommunications operators by market cap by mid-2015).[6] It has already been stated that M2M connections constitute a small percentage of total mobile connections (only 3%), and a similar figure (2%) of total cellular traffic.[7]

Besides published data from operators, some national regulators have disclosed M2M data for their countries. For example, the national regulatory authority in Brazil (Anatel) discloses M2M results monthly, for example, for May 2015 (rounded to the nearest one hundred thousand for subscribers and to the nearest percentage integer for market share),[8] as shown in Table 6.2.

In Spain, its regulator (*Comisión Nacional de los Mercados y la Competencia-CNMC*) just publishes the total,[9] 1.7 million lines by February 2015.

6.2.2 Pricing

In terms of pricing, an example taken from the figures publicly listed[10] by MyM2M indicated the pricing for a global M2M player (Wyless) that partners with mobile network operators (Vodafone, Telefonica, Verizon, Singtel, America Movil, AT&T, Deutsche Telekom/T-Mobile, Softbank/Sprint, Rogers, Telus, etc.):

- SIM: €5 (including same amount as balance, free shipment for 10 or more)
- Monthly recurring charge: €1/2/3 (low/medium/high usage)
- Usage price per megabyte (MB) €20/10/7.5 cents, respectively, in Europe, €45 cents in the world

This means that a price per megabyte can be as low as €7.5 cents.

Table 6.2 Machine-to-Machine Connections and Market Share by Telecom Operators in Brazilian Market

Operator	Subscribers (million)	Market share
America Movil (Claro)	4.1	39%
Telefonica (Vivo)	3.9	37%
Telecom Italia (TIM)	1.4	13%
Oi	1.2	11%
Total	10.6	

6.2.3 Networks

From a pure connectivity perspective, there are different options in terms of the underlying cellular technology. The three main ones correspond to the evolution of cellular data: 2G (GPRS or equivalent in CDMA technology), 3G (UMTS/HSPA or CDMA equivalent), and 4G (LTE). Of course, the main difference among them is about speed (and capacity), with a trade-off with coverage, which is not yet universal for 4G, not even for 3G.

But also the potential decommission of legacy networks is a key decision factor, given the typical long (many years) timeframes for M2M deployments. In countries such as Japan, 2G was already switched off a few years ago (NTT Docomo in 2012,[11] Softbank in 2010,[12] and KDDI in 2008).[13] Plans for shutting down legacy networks are more advanced among CDMA operators, but there are also some examples among GSM operators, for example, a major one is AT&T, [14] which announced a long time ago its intention to completely abandon 2G by 2016. As a matter of fact, it is actively progressing toward the goal, by refarming 2G spectrum for LTE (i.e., transmitting LTE in spectrum that was formerly used for 2G), with most markets already using 5 or 10 MHz of spectrum for 2G. Another example with a similar date is Telstra in Australia. An alternative choice is to shut down 3G first, which is planned by telecommunications operator Telenor[15] in Norway for 2020, 5 years ahead of 2G decommissioning, mainly due to M2M-installed base that is mostly 2G. If operators shut down legacy networks, they have to migrate the M2M-installed base out of those networks. They may need to replace the communications modules of the endpoints.

Finally, the chipset/module cost is a variable (around $10[16] for 2G, with a decreasing trend as volumes grow). The newer technologies are more expensive, although the gap is decreasing as they mature and reach scale. Of course, the modules can include all technologies but they will become more expensive. For price and coverage reasons 2G modules were still dominant (more than 50%) in terms of volume in 2014 (although decreasing from 70% in 2013).[17] In terms of revenues, $1 billion has already been exceeded in 2012 (and in 2013). The main module vendors are Sierra Wireless (market leader in both revenues—for four consecutive years, estimated 34% by mid-2014[18]—and shipments,[19] having shipped more than 100 million), SIMCom Wireless (Chinese), Gemalto, and Telit Wireless Solutions.

6.2.4 Managed Connectivity

On the other hand, taking into account scenarios with a huge number of lines, some sort of management portal is required to enable easier operation. That is offered by managed connectivity, which is available as an additional layer on top of individual connectivity, and is paid for by customers. Two main platforms come from Jasper Technologies (a specialized M2M new entrant) and Ericsson (a traditional telecommunications vendor). Most telecommunications operators offer one of the two commercial platforms, although there are some notable exceptions in Table 6.1, and sometimes commercial platforms coexist with internally developed platforms, and are even referred to as partners by both Jasper and Ericsson. End customers also require these platforms when awarding large contracts.

Typically managed connectivity platforms are offered as a service (e.g., at the cloud), and as mentioned they simplify the management processes by enterprise customers. Functionalities include management of subscriptions, rating and billing, provisioning and order management, connectivity monitoring, security, traffic, notifications, reports, statistics, and so on. Device management is also possible.

Jasper has partnerships with operators such as AT&T in the United States; Bell, Telus, and Rogers in Canada; America Movil, Telefonica, Millicom, and Telecom Personal in Latin America; Telenor, KPN, Telecom Italia, Tele2, Vimpelcom, and Telefonica in Europe; Singtel, Etisalat, NTT Docomo, and China Unicom in Asia Pacific. Most of them[20] belong to M2M World Alliance. Rogers, Telefonica, KPN, Vimpelcom, Singtel, NTT Docomo, and Telstra were founding members in 2012, while Etisalat and Telenor joined in 2013 and 2014, respectively. M2M World Alliance was founded to offer global M2M jointly to multinational customers, given the fragmented footprints of mobile operators. The Alliance offers a single global SIM and a management portal.

Ericsson offers the *Device Connection Platform* (DCP), which builds on the acquisition in 2011 of a Telenor platform. It has signed an agreement with *Global M2M Association* (Deutsche Telekom, Orange, TeliaSonera, Telecom Italia Mobile, Bell Canada, and SoftBank) and *Bridge Alliance* (mainly Asia-Pacific but also Middle East and Africa, 36 operators including Airtel, SK Telekom, Globe, Telkomsel, and Softbank).

6.2.5 Value Chain

Before finishing this section, let us show in Table 6.3 a simplified value chain of M2M connectivity business, indicating a few examples of major players in each of the links, which have already been mentioned.

Table 6.3 Machine-to-Machine Sample Players in the Value Chain

Radio chipset	Qualcomm
Smartcard (SIM)	Gemalto, G&D, Oberthur
Communication module	Sierra Wireless, SIMCom, Gemalto, Telit
Management portal	Jasper, Ericsson
Network	China Mobile, AT&T, Vodafone, Telefonica, Deutsche Telekom, Telstra
Customer	General Motors, etc.

6.2.6 Embedded SIM and Remote Provisioning

As opposed to the consumer space, in which the user can change the SIM if needed, or if not already inserted in the device (e.g., if it has been bought in the operator retail channel), for example, in the case of an unlocked device, the user can insert it the very first time the device is used. But for M2M the typical scenario is that the communication module (and therefore the SIM) is included in the manufacturing process of the machine itself (e.g., a car or a smart meter mentioned in Section 6.1). For these typical cases, it is cumbersome to include a SIM that has been specifically delivered by a certain operator with which the connectivity contract has been signed. It would be more convenient to provide a generic SIM that is provisioned with operator credentials dynamically when the device connectivity needs to be activated.

That is precisely the purpose of the so-called embedded SIM, also known as eUICC (Embedded Universal Integrated Circuit Card). Backed by GSMA (GSM Association: the world association of mobile telecommunications operator), there is a standard mechanism that allows an *"over-the-air* capability" (i.e., using the data channel) to transfer the credential of the initial operator, and even to a switched operator later on. This is also useful for multinational contracts, in which several operators (see the alliances in Section 6.2.4) provide the service to a single customer, depending on the country where the machine is installed. Moreover, it would also allow a customer to change the provider once the contract period finishes. Again, think of scenarios where changing the SIM is very complex (remote locations, difficulty in accessing it within the machine of communication module, etc.) or not feasible at all. Of course, a key feature is to maintain the highest security level offered by SIMs, which should not be compromised because of remote transmission. One of the very first industries that has backed embedded SIM is automotive, somehow a bellwether for its adoption. Support (deployment or commitment) by operators is massive, with 20 of them accounting for roughly two thirds of world connections by early 2015,[21] given the support by two main M2M alliances (already mentioned in Section 6.2.4: M2M World Alliance and Global M2M Alliance), plus major operators such as China Mobile (first in the world ranking by number of M2M connections), America Movil, and so on. In terms of costs, GSMA-embedded SIM specifications have been calculated to be 90% cheaper than alternative proprietary solutions.[22]

On June 30, 2015, GSMA published *Remote Provisioning Architecture for Embedded UICC Technical Specification Version 3.0.* [23]

6.3 LOW-POWER WIDE AREA: LTE-MTC AND ALTERNATIVES

It has already been stated that machines have very different connectivity needs compared to people. This applies to data volumes, speed, and frequency of usage. A long tail of devices (and even sensors) cannot afford expensive communications modules, which cost several dollars or tens of dollars for normal wireless devices, much higher than emerging technologies with just $1 Bill of Materials incremental cost.[24] They cannot always connect to electricity grid and therefore require minimum power consumption in order to run on batteries for years, and are not served by cellular coverage (very remote or underground areas with difficult propagation). In order to cope with this long tail, which could be in the billions order of magnitude, alternative radio technologies have to be designed. They constitute Low-Power Wide-Area (LPWA) technologies, which can be divided in two groups. The first group is pursued by several start-ups with proprietary solutions. The second group pursues the evolution of wireless standards, namely, new releases of 3GPP, called LTE Machine-Type Communications (LTE-MTC or simply LTE Machine, LTE-M), but also WiFi evolution (802.11ah).

In the first group, there are companies such as Sigfox, Neul, Actility, and On-Ramp (renamed to Ingenu), which are attracting large investment sums and some of which are working around pursuing a standard and its associated ecosystem. They are covered in the following sections, although comparing them is difficult since not all have reached the same commercial maturity and some of their claims are only theoretically shown in their specifications.

The list is much larger, including others not covered in the following sections[25]:

- *Amber Wireless* [26]: Wireless metering adapters with wireless M-Bus (169 and 868 MHz), low power WiFi module.
- *Telensa* [27] (which in March 2015 integrated Senaptic that was found in 2014[28]; both formed by technology provider Plextek[29]): Also based in Ultra-Narrowband (UNB), claims 8 million active devices in 30 countries including the United Kingdom (10% of UK street lights) and Brazil.
- *Elster*[25] *(which acquired Coronis several years ago):* Focus on water metering.
- *M2M Spectrum Networks* [30]: Licensed spectrum 800–900 MHz, U.S. nationwide deployment.

6.3.1 Sigfox

Sigfox is a French start-up, which raised $115 million[31] in an investment around early 2015 that included mobile network operators NTT Docomo, SK Telekom, and Telefonica. It is probably the most advanced in terms of deployment (more than 2 million km^2 already covered, out of a world land total of approximately 150), starting with Europe—typically, with cell towers/TV infrastructure operators such as Cellnex in Spain ($5 million investment over 7 months[32]), or Arqiva in 10 cities in the United Kingdom or self-deployment by Sigfox such as that in France (or announced in the United States?). Other countries deployed by mid-2015 include Portugal and The Netherlands (85% coverage), followed by announcements for Denmark and Belgium (with deployment over 2 years by utility Engie, formerly known as GDF Suez[33]); and cities such as Warsaw (Poland), Munich (Germany), Milano (Italy), Dublin (Ireland), Prague (Czech Republic), and Stockholm (Sweden) in Europe; Santiago (Chile), Medellin (Colombia), and San Francisco (the United States) in Americas; and Mumbai (also known as Bombay, India) in Asia.[34]

Also, from a usage perspective, Sigfox is relatively advanced, claiming to have reached a quarter of a billion (cumulative) messages sent in its network by June/July 2015. These messages were generated by a moderately small number of objects, around 300,000 devices by March 2015. Pricing starts at $8 per year, decreasing with volumes.[35]

Chipset and module manufacturers that have partnered with Sigfox include Texas Instruments, Telit, STMicroelectronics, Intel, and so on.[36] In the 2015 investment round press release the company claims[35] its complementarity to cellular networks, and the future integration with them. In the following paragraphs, the main technical features (power, coverage, throughput, spectrum, etc.) will be described[37]:

In terms of spectrum, it is based on unlicensed (ISM that stands for Industrial, Scientific and Medical) band at 868 MHz in Europe and 902 MHz in the United States, but indicates that it can adapt to other frequencies.

Based on ultra-narrowband, the low-throughput devices can *send* (uplink) up to *140 messages per day*, each of which can be up to *12 bytes* (data payload). Downlink devices can receive *four messages of 8 bytes* (payload) *per day*. But before doing so, the device has to request it, meaning that it has to be programmed to *receive data at specific events or at specific times*. In terms of power, typical emission consumes between 20 and 70 mA, almost nonconsuming when idle, with unidirectional communication (transmit only) to minimize power consumption. Emission power is between −20 and +20 dBm, being a trade-off between range and power consumption. The radiated power is limited by regulations of ISM bands (below 25 mW).

In terms of coverage or range, typical cell deployment is planned between 30 and 50 km in rural environments, but the distance is reduced to 3–10 km in urban areas. For line-of-sight applications, the distance could exceed 1000 km. In terms of capacity, each base station can handle 1 million objects connected.

A summary is provided in Table 6.4.

6.3.2 LTE-MTC

The LTE Machine-Type Communications is the name given to the evolution of cellular standards in order to cope with the requirements imposed by machine

Table 6.4 Sigfox Main Features

	Spectrum	Throughput	Operators
Sigfox	Unlicensed 900 MHz	1680 bytes/day uplink, 32 bytes/day downlink	SK Telekom, NTT Docomo, Telefonica

communications. Besides coverage, power consumption and low cost, typical features are low (if any) mobility, potentially huge number of connected devices, low traffic volume per device, generally uplink traffic, can cope with constrained time windows for reception (or even transmission) of data, security of traffic, and so on.

The feature for Machine-Type Communications was started for consideration in Release 11 (e.g., user equipment power preference indication or Radio Access Network overload control). But the first major set of improvements appeared in *Release 12 of 3GPP* (Third-Generation Partnership Project), which has been frozen in *March 2015*. Communication modules conformant with this release are called *LTE Category 0*. The main difference with previous LTE releases is related to reducing power consumption, so that devices can run longer on a battery (without being plugged to the electricity grid, which might not be available). *Release 13* is expected to be frozen in *March 2016*, including even less power consumption, reduced costs and coverage improvements (hard to reach sites), and some call-associated devices Category-1.

LTE User Equipment Category 1 devices have a maximum downlink speed of 10 Mbps, compared to 1 Mbps of Category 0. In 2014 (last quarter), manufacturer Sequans together with operator Verizon and equipment vendor Ericsson successfully tested Category 1 devices,[38] with a similar trial with another major U.S. operator in first quarter 2015.[39] As opposed to Category 0, which requires network upgrades, Category 1 could readily be used, since it was already specified in the earliest LTE releases. The cost of modules can drop below those of the 3G modules. Note that release 13 could not be available generally until 2018,[35] and it has to be confirmed whether Release 12 would be available commercially in 2016.[40]

A chipset manufacturer (Altair) already announced in February 2015 the availability of samples of Category 1 and Category 0 devices,[41] allowing commercial applications to start by the end 2015. They are software upgradeable to support Release 13 features. They claim to have very large demand, signing several contracts even before the sample testing was done, which seems to be unusual in the industry. Another claim is to improve power management beyond Release 12/Category 0 specifications, maximizing sleep time, so that a single battery can last for up to 10 years.[42]

A good reference summarizing all the LTE improvements for M2M is *Recent advancements in M2M communications in 4G networks and evolution toward 5G*,[43] which shows that the modem complexity relative to Category 1 is 50% of Category 0 (Release 12), and 25% for Release 13 low-cost. Another good reference is a white paper by telecommunications vendor Nokia.[44]

3GPP references provide additional detailed information:

- Technical Specification Group Services and System Aspects, including the following documents:
  - *Service requirements for Machine-Type Communications; Stage 1*, labeled *3GPP Technical Specification TS 22.368* [45]: Release 12 version 12.4.0 available in June 2014, corresponding to ETSI (European Telecommunications Standards Institute) Technical Specification TS 122 368 V12.4.0

published in October 2014; Release 13 version 13.1.0 available December 2014.

    – *Study on Machine-Type Communications and other mobile data applications communications enhancements (Release 12)* [46]: Technical Report TR 23.887 V12.0.0 (December 2013).

- Technical Specification Group Radio Access Network provides the following:
  - *Study on Enhancements to Machine-Type Communications and Other Mobile Data Applications; Radio Access Network (RAN) aspects (Release 12)* [47]: 3GPP Technical Report TR 37.869 V12.0.0 (September 2013), which deals with topics of signaling overhead reduction and user equipment power consumption optimization.
  - *Study on provision of low-cost Machine-Type Communications User Equipment (UE) based on LTE* (Release 12), 3GPP TR 36.888 V12.0.0 (June 2013)[48]: Cost analysis methodology is based on baseband cost/complexity and Radio Frequency (RF) cost analysis. Cost improvements are based on reduction of maximum bandwidth (downlink bandwidth decreased to 1.4 MHz), single receive RF chain, reduced peak rate (reduction of maximum transport block sizes, restricted to 1000 bits), reduced transmit power (with a trade-off with coverage extension), half duplex operation (as opposed to full duplex), and less supported downlink transmission modes. The same document also devotes a chapter to coverage improvement, with a target gain of 20 dB (needless to say this a logarithmic scale, 10 times logarithm in base 10, that is, 100 times), but considering trade-offs, it could be 15 dB at least for FDD (Frequency Division Duplexing); a mechanism to inform the coverage needs of the user equipment is also included. It is important to note that changes proposed do not require a hardware upgrade on the network side (eNodeB, short name for evolved UMTS Terrestrial Radio Access NodeB that interfaces the User Equipment). The Bill of Materials (BOM) cost of User Equipment modem would be reduced to a similar order of magnitude to 2G EGPRS (Enhanced General Packet Radio Service, also known as EDGE: Enhanced Data rates for Global Evolution).
  - Evolved Universal Terrestrial Radio Access (E-UTRA); User Equipment (UE) radio access capabilities (Release 12),[49] 3GPP TS 36.306 V12.4.0, corresponding to ETSI TS 136 306 V12.4.0 (2015-04). Related to LTE User Equipment categories, 3GPP[50] references this interesting article about LTE UE Category and Class Definitions.[51]

A summary is provided in Table 6.5.

Before finishing this section, let us mention the following two additional cellular technologies:

- *EC-GSM:* Extended Coverage GSM, which as its name implies aims to stretch 2G technology by improving its coverage.

Table 6.5   Long-Term Evolution-Machine-Type Communications Main Features

3GPP release/LTE UE category	Speed (Mbps) (downlink/ uplink)	Half duplex (optional)	Bandwidth (MHz)	Maximum transmission power (dBm)	Expected availability
8/Category 1	10/5	No	20	23	2015
12/Category 0 (Cat-M or Cat-M1)	1/1	Yes	20	23	2016
13	1/1	Yes	1.4	20	2018

Note: dBm = $10^*\log_{10}$ power/1 mW. For example, 20 dBm means 0.1 W.

- *LTE-CAT M2* (formerly known as NB-IoT—narrowband IoT), which is a complementary cellular technology (40 kbps throughput) in order to match the features of noncellular technologies, which are difficult to be replicated with the traditional evolution of cellular standards. It is being included in Release 13.

6.3.3   Weightless

The Weightless standard is advocated by its Special Interest Group (SIG).[52] Version 1 of the *Weightless-N* standard was published as an open standard (as opposed to proprietary Sigfox, which is also royalty free for device vendors) in *May 2015*,[53] although not publicly available but only for their more than 2000 members worldwide. The main features displayed on their website are 5 km range (urban, 20–30 km rural; coping with a typical 15–20 dB building penetration loss),[54] 10-year battery life (single AA battery), $2 module cost (even $1 chip in volume given commodity components such as transceiver, microcontroller, regulator, crystal, few passives and cheap antenna, and $3000 base station Bill of Materials), and a few dollars per year subscription fee (low traffic volume). Like Sigfox, it operates on sub-1 GHz unlicensed ISM spectrum (pivoting from Weightless-W that was thought for TV white space spectrum, which is subject to regional licensing limitations, being available only in 400–800 MHz range in a few countries such as the United States, Canada, and Singapore,[55] expectations to regulatory availability extension to other countries has not happened as expected[56]); trade-off is speed reduction, from tens of kilobits per second to tens of bits per second, given typical worse propagation in higher frequencies, and is based on UNB technology. It allows several networks to coexist (whereas Sigfox is based on agreements with a group called Sigfox Network Operators). Initial version is limited to uplink communications, although bidirectionality is foreseen in the future. nWave Technologies joined the SIG in October 2014, and as a core member has contributed significantly to the development of the standard. Just 1 month earlier, Weightless-N standard had been announced.

In June 2015, the deployment of a network in London was announced, based on Weightless-N operation of nWave technology (which[57] claims 10-year duration

Table 6.6 Weightless-N Main Features

	Spectrum	Range	Battery duration	Module cost	Operators	Commercial
Weightless-N	Unlicensed 900 MHz	5 km urban, 20–30 km rural	10 years	$1–2 in volume claim	BT	Only very small-scale trial by mid-2015

with 2.5 mAh battery, low cost of modem, 10 km range). However, besides this, it has to be noted that detailed information about hardware (including development kits) by Special Interest Group was not yet available as of mid-2015,[58] showing less commercial maturity than Sigfox. Neul had produced the first chipset for the proposed Weightless standard (by then new N variant had just been announced a few days earlier), but the company was acquired by telecom equipment vendor Huawei in September 2014, as we will see in next section. BT (also based in the United Kingdom such as Neul) had adopted Neul Weightless technology for IoT trials in January 2014.[59]

The Weightless-N main features are summarized in Table 6.6.

6.3.4 Cellular Internet of Things (Huawei)

Besides the Weightless standard, there is another standard called Cellular Internet of Things (CIoT). It is backed by major telecommunications equipment vendor (Huawei) that acquired a start-up (Neul) for about $25 million in 2014 (as it has been just commented). But it is worth highlighting that its aim is to standardize (open industry standard) Cellular IoT as part of the cellular track (3GPP, Third-Generation Partnership Project standards). With Neul, Huawei has reinforced its position as a telecommunications vendor for IoT, acquiring a company with working chipsets as well as intellectual property. In Mobile World Congress 2015 (March), Huawei already showcased what they called precommercial LTE-M in a smartband.[60]

The first telecommunications operator announcing its support has been Vodafone.[61] It builds on existing cellular/M2M networks and provides a dedicated network access layer specifically designed to support Low-Power Wide-Area devices.

One example of Huawei's contribution to 3GPP is the document *CIoT - Coexistence with GSM*,[62] which tries to deal with the avoidance of negative impacts to legacy cellular systems in the same frequency band, providing a common evaluation framework.

6.3.5 LORA (Long Range)

LoRa Alliance is another open, nonprofit association around LPWA that proposes the LoRaWan protocol.[63] Founded in January 2015, it comprises members of different links of the value chain, including operators (such as Bouygues Telecom in France, Proximus/Belgacom in Belgium, KPN in The Netherlands, Swisscom in Switzerland), equipment vendors (such as Cisco), semiconductors (such as Semtech—analog and mixed-signal products—and one of the main developers of the technology), sensor manufacturers, and systems integrators (such as IBM). From a technology perspective, it provides *bidirectional communications*. The network topology is based on *intermediate gateways* between devices and backend. The gateways provide Internet Protocol connectivity to the central server. *Data rates are adaptive* in order to maximize battery duration using its Adaptive Data Rate algorithm, ranging from 0.3 to 50 kbps. The selection of data rate and frequency channel is a trade-off between range and capacity. It is designed for devices that can run on a coin cell battery. It also provides security functions.

The following are three types of devices (labeled Class A/B/C), all of them bidirectional:

- Class A ones are the most power efficient, only allowing reception of data immediately after transmission (two short reception windows).
- Class B have scheduled receive slots in addition to those of Class A. At schedule times the gateway sends a beacon to the device allowing the server to know when the device is listening.
- Class C has the most receive slots, since they have nearly continuous receive windows.

It also allows multicasting, enabling remote over-the-air software upgrades.

LoRaWan specifications have been released in mid-2015 (June)[64] and can be obtained upon request (e.g., by developers). As part of the ecosystem, it has a certification program (LoRa Alliance Certified Product and LoRa Alliance Compliant Platform).

It claims to have advantages over competing technologies, such as bidirectionality, security, mobility, and accurate location capabilities.

In June 2015, the start-up Actility raised $25m[65] in a founding round led by three telecommunications operators (KPN, Swisscom, and Orange) and manufacturer Foxconn. Actility claims to provide the first end to end LoRaWan interoperable platform over unlicensed ISM band spectrum and the ability to support commercial networks in less than a month, with the capabilities of customer enrolment, activation, and billing. It also complies with ETSI M2M standard supported by OneM2M (which released their first standard specification in February 2015).[66]

In terms of commercial deployments, Bouygues announced in March 2015 the launch of a LoRa network in France in June following a successful 16-month trial in a city, in close collaboration with Semtech. Starting with Paris with a target to cover 500 towns by the end of 2015, Bouygues has thus become the very first telecommunications operator to deploy a LPWA network in France (taking into account the Sigfox deployment in the country with no operator involvement).[67]

The features of LoRa are summarized in Table 6.7.

Table 6.7 Long-Range Main Features

	Battery duration	Throughput	Operators	Commercial
LoRa	10 years (single-coin cell)	0.3–50 kbps	Bouygues, Proximus, KPN, Swisscom, Orange	Bouygues mid-2015 in France

6.3.6 WIFI: 802.11ah

WiFi is a Local Area Network (LAN) technology, and therefore its range is much shorter than Wide Area Networks (WAN) covered in previous sections. However, with *ah* release, its range could be extended to the order of magnitude of a kilometer (more modest than alternatives), based on sub-1 GHz (also known as S1G) operation (with 1/2 MHz bandwidth (?) typically but also 4/8/ 16 MHz modes, the higher only in those countries with more bandwidth available such as the United States or China[68]), and speeds higher than 100 kbps, or even much higher. Power consumption is also optimized (e.g., deactivating radio in periods without activity, shorter frames/packets, etc.) for *battery duration of years*, as well as scalability (several thousand devices per access point). One cannot ignore the fact that, like cellular, WiFi enjoys economies of scale as a globally established standard.

The task group with this name was established in 2010. Final approval is expected in early 2016.[69] Status can be seen in the task group portal.[70] By the end of June 2015, the candidate version is in draft 5.0, dated March 2015.[71]

Good reference papers about the topic are *IEEE 802.11AH: The WiFi Approach for M2M Communications*[72] and *Outdoor Long-Range WLANs: A Lesson for IEEE 802.11ah.*[73]

The physical layer is based on Orthogonal Frequency Division Multiplexing (OFDM) (based on 802.11ac modern WiFi standard that operates on 5 GHz, but optimized for frequencies below 1 GHz) with enhancements in the Medium Access Control (MAC) layer. Like most other alternatives, it works on unlicensed bands below 1 GHz excluding TV white space bands, which depend on the country/region (besides already commented Europe -863/868.6 MHz and U.S. -902-928 MHz-; Japan 915–929 MHz, South Korea 917-923.5 MHz, Singapore 866-9 and 920-5, and China 614-787 MHz or 779-787).

The features of 802.11ah WiFi is summarized in Table 6.8.

Table 6.8 802.11ah Main Features

	Spectrum	Throughput	Range
802.11ah	<1 GHz unlicensed	>10 kbps	Kilometer

6.3.7   On-Ramp

Let us not finish this survey of alternative Low-Power Wide-Area technologies without mentioning On-Ramp Wireless. It is based on RPMA® (Random Phase Multiple Access) technology, which provides secured bidirectional communications, and as usual in this space operates in the unlicensed spectrum, but as opposed to others, in 2.4 GHz (which apparently has worse propagation features). By the end of 2014, it had deployed 30 networks[74] mainly in the *energy sector*, where most of its investors come from. The latest round was a series C of $31 million in 2013, which includes existing investors such as GE Ventures through a joint venture. Some of the latest additions by mid-2015 include Italy, where a partner is deploying a network for natural gas smart metering (and water in the next stage), a partnership for smart metering in Asia Pacific, and in the United States where a partner is providing coverage for over 55,000 square miles of active oil and gas fields, representing more than half of total U.S. production, and supporting 5000 radios per access point.

Regarding standardization, the company was a founding member of IEEE 802.15.4k Low-Energy Critical Infrastructure Monitoring task group. 802.15 deals with Wireless Personal Area Networks, and is the basis for ZigBee.

The company claims that its physical layer design is able to reach up to 172 dB *of path loss*. A single access point can support tens of thousands of end points concurrently. The battery duration is also of several years, such as competing alternatives.

The company's products include a radio module with small form factor that can be integrated surface-mounted, through an industry standard Serial Peripheral Interface. They also provide a reference platform for third parties developing sensor applications.

6.4   APPLICATIONS: TOWARD INTERNET OF THINGS

According to industry analysts such as Machina Research, there were already 5 billion connected objects in 2014, referring to any object with any kind of connectivity including short range. As a matter of fact, most of them (72%) are short range as in Local Area Networks such as WiFi, Ethernet, or Power Line Communications; or even smaller personal area networks such as Bluetooth or ZigBee. They are predominantly within buildings including consumer electronics devices.[14] This figure dwarfs machine cellular connections (but not all cellular connections, machine and human, see Chapter 2), which, as already stated, were just a quarter of a billion by 2014, which is just 5% of the total. In terms of revenues, the total is half a billion U.S. dollars in 2014, including devices (which constitutes the highest share), connectivity (which in some cases is free due to the use of unlicensed wireless bands), and application. Without aiming to distract the reader with terminology issues, Internet of Things is a wider concept, including additional elements of the value chain beyond device, connectivity, and applications: platforms and middleware, data monetization and IoT services, installation, M2M services, systems integration, and strategy. Taking all these elements into account, the figure is even larger, $900 billion revenues for Internet of Things[75] in 2014. In the long term, IoT will be the convergence of M2M and Information Technology sectors.

In terms of major players, IBM (Information Technology sector) announced in March 2015 that it will invest $3 billion over 4 years in its new IoT unit that has 2000 people, and leveraging cloud open platform, big data analytics, and developer ecosystem.[76] Intel (semiconductors sector) reported $2.1 billion revenues in 2014 (19% year-over-year growth, with $616 million operating income) for the Internet of Things business unit (out of nine operating segments, the one with largest growth, third in terms of revenues in fourth quarter of 2014, and representing almost 4% of total revenues in the year). The unit was created in the first quarter of 2014 and includes platforms and software for embedded markets such as retail, transportation, industrial, buildings, and home, with record full-year unit shipments.[77] Qualcomm (also in the semiconductor industry, more focused on mobile), surpassed $1 billion revenues in 2014 (chipsets only as opposed to Intel), and expects to generate more than 10% of its revenues in 2015[78] (around $2.6 billion according to guidance[79]). In addition to smart homes (connecting 120 million devices in 2014) and wearables (20 different devices), which are covered in other chapters related to consumer segment, it powers 20 million connected vehicles from more than 15 car manufacturers, involved in more than 20 smart cities projects (water, recycling, lighting, transportation, energy, and infrastructure), health care (about 500 customers, offering connected health solutions, for example, for chronic disease management).[80] For General Electric, an industrial conglomerate, its Industrial Internet revenues exceeded $1 billion ($1.3 billion in orders) in incremental revenue in 2014 (*Predictivity* solutions with more than 40 offerings). It opened up the software platform called *Predix* that powers its *Predictivity* solutions to third parties in 2015.[81] GE has 1000 programmers and data scientists,[82] thanks to an investment of $1 billion in its Industrial Internet.[83] Partnerships include Intel (architecture for integrating devices), Cisco (integrating networking equipment in the management of assets in industrial locations), and telecommunications operators (for global connectivity), including AT&T, Softbank, Vodafone, and Verizon. It claims to analyze daily 50 million data elements from 10 million sensors on $1 trillion of managed assets (asset performance management).

6.4.1 Forecasts

There is a lot of hype in forecasts, and although they are only speculative, let us show what analysts expect for the long term in Table 6.9.

6.4.2 Connected Car

As already commented, automotive represents a major vertical for connected objects. It has to be noted that Fleet Management for enterprises is probably the

Table 6.9 Internet of Things Forecasts

Forecast/analyst	Machina 2024[a]	IDC 2020[b]	Gartner 2020[c]
Connections (Devices)	27 billion (18% Compound Annual Growth Rate (CAGR))		25 billion
Connections>Technology>Short range	69% (decrease of 3 percentual points)		
Connections>Technology>Cellular	2.2 billion (>50% connected car)		
Connections>Technology>Low-Power Wide Area	14%		
Connections>Market>China	21% (15% of cellular)		
Connections>Market>United States	20% (20% of cellular)		
Connections>Market>Japan	8% (6% of cellular)		
Traffic>Cellular	4% (increase of 2 percentage points)		
Revenues	$1.6 trillion (12% CAGR) M2M/$4.3 trillion IoT	$1.7 trillion (17% CAGR)	$263 billion (services)
Revenues>Markets> United States	23%		
Revenues>Markets>China	19%		
Revenues>Markets>Japan	7%		
Revenues>Categories>Devices+Connectivity+IT services		Over 67%	
Revenues>Categories>Devices (modules/sensors)		32%	

[a] Last 2G phone shipped 8 months ago in Japan (*KDDI/AU switched off their 2G radio network in March this year*). Eurotechnology Japan KK ©2013, //www.eurotechnology.com/2008/08/22/last-2g-phone-shipped-8-months-ago-in-japan/ (August 22, 2008).

[b] Explosive Internet of Things spending to reach $1.7 trillion in 2020, according to IDC. Copyright 2015, IDC, //www.idc.com/getdoc.jsp?containerId=prUS25658015 (June 2, 2015).

[c] Gartner says 4.9 billion Connected "Things" will be in use in 2015. © 2015 Gartner, Inc., //www.gartner.com/newsroom/id/2905717 (November 11, 2014).

oldest (pre-)M2M service in the market in any vertical that existed with proprietary solutions even before M2M communications were available and being offered by niche providers. For enterprises with several vehicles, the customer value proposition is around optimizing operations (reducing fuel consumption, location of the different vehicles, etc.).

In this section, the automotive vertical will be addressed. It is typically referred to as Connected Car. But before starting, let us make a distinction between two main application types, infotainment (information and entertainment) for people versus more car-related telematics.

Regarding the former, in the current world of massive smartphone adoption, it is quite natural to leverage smartphone capabilities for infotainment also within the car. Typically, smartphone connects to the car (initially wired, e.g., USB—and now expanding to wireless), leveraging its user-interface capabilities, namely, the dashboard for visual display, speakers and microphones for audio, controls (steering wheel, touchscreen, knobs, buttons, dials, etc.) as input, GPS for positioning, and so on. Complementarily, voice command recognition as well as text to speech functionality are also available and their uses are encouraged. This is quite convenient for the driver, avoiding distractions that can be fatal. At the same time, it allows the motorists to use the same content and associated apps that are already available outside the car through the smartphone. Among applications, one that stands out naturally is the navigation map for driving. Another interesting use case is associated with shared cars (e.g., rental cars or even the future trend of sharing cars directly among people), since infotainment preferences are portable among cars.

The key requirement for this to happen is to enable compatibility both in the car and in the smartphone. One of the first platforms to make this happen has been *MirrorLink* (by Car Connectivity Consortium). Detailed information about compatible cars, smartphones, apps, and audio systems can be found on its website.[84] However, the availability (2014 and 2015) by smartphone OS behemoths Apple iOS and Google Android, together with their own platforms, Apple *CarPlay* and *Android Auto* (fostered through Open Automotive Alliance), could accelerate its massive adoption by car manufacturers. Many of them are announcing support for both platforms, giving the end customer freedom to use in the car their smartphone of choice. In many cases, the announcements are for new car models; however, existing ones could be somehow upgraded through external systems (aftermarket), particularly in the audio from major manufacturers. The list of supported countries (typically starting with major ones) and vehicle manufacturers is large and growing. Again further information (e.g., supported operating systems versions) can be found on their respective websites.[85,86,87,88]

On the other hand, infotainment can be extended to other devices from passengers with no cellular connectivity. A typical way of providing connectivity is through in-car WiFi, powered by an access point within the car that in turn connects to a cellular network (typically 4G LTE if available).

Besides infotainment, cars themselves can benefit from connectivity. Just think of car diagnostics, safety (e.g., emergency call), security (stolen vehicle recovery, or even plain vehicle location in case the driver has forgotten), and comfort (turning on heating, ventilation, or air conditioning a few minutes before going to the car). All these features cannot rely properly on an external device such as a smartphone (which is either not robust enough, or simply is not always inside the car).

OnStar by General Motors is probably the most prominent service. By mid-2015 it was offered in the United States in three plans at a pricing of $19.99, $24.99, and $34.99 per month or per year (the latter with discount). However, it has to be noted that there are promotions offering the service for free typically for the first year of a new car, and even 5 years for most basic features related to vehicle diagnostics (even proactively warning of potential failures before they happen, e.g., the battery, starter, fuel pump), dealer maintenance notifications, and remote operations of certain car features.[89] The basic plan covers safety and assistance, and is useful for emergency situations, offering automatic crash response, roadside assistance, and permanent access to advisors. The middle plan additionally includes security antitheft measures such as stolen vehicle assistance (with remote ignition block to prevent starting the engine and vehicle slowdown in case the vehicle is moving) as well as theft alarm remote notification. Finally, the most expensive plan is less related to car telematics and offers convenience services for the people, such as remote operation of the car (directly or through call center: honk horn, flash lights, start vehicle, lock/unlock doors), navigation, hands-free calling minutes, and some sort of concierge service, for example, for reservations related to the trip (restaurant, hotel, shopping, etc.). Besides communications add-ons (voice and data plans), it offers a $3.99 bolt-on for car location, designed for peace of mind. Moreover, several auto insurance companies offer discounts based on low mileage or OnStar theft prevention features[90]; but in 2015, a more advanced discount offering was announced, including a partnership with a first (among the ones just mentioned) insurance company.[91] These kinds of services assess the driving behavior and style, based on data such as speed, acceleration, hard stops, and rural versus urban roads. They in turn provide useful feedback for safer and cheaper driving based on benchmarking with other drivers. And naturally, auto insurance premium can be customized, if the customer grants permission to access their personal data, given the more accurate (compared to traditional means such as demographics information) risk profile of the driver. Other relevant feature is the remote update of software, over-the-air term used in mobile industry. Full details about the OnStar service can be checked on its website.[92] The service has 7 million subscribers in the United States (where it started around 20 years ago), Canada, China, and Mexico (since 2013).[93] In 2015, it is expanding into Europe (through its subsidiary Opel; 13 countries including the largest ones, with WiFi feature starting only in a few; full details on the website)[94,95] and Brazil.[2] In early 2012 (when it had 6 million subscribers), it reached an agreement with telecommunications operator Telefonica, for its international expansion (outside United States, Canada, and China).[96] Financial details (profit and loss) have not been traditionally disclosed, but some hints can be found either by sporadical comments by the company or hints by industry analysts (e.g., in this chapter).[97] Finally, it is worth mentioning that OnStar is also offered for vehicles from other manufacturers, through the aftermarket service *For*

My Vehicle, which is based on replacing the rearview mirror; for additional details, see its website.[98]

A couple of telecommunications operators have entered the space through acquisitions. In 2012, Verizon acquired, for a total of $612 million, Hughes Telematics Inc,[99] which after integration became Verizon Telematics. Given the size of the transaction, it is likely one of the largest parts of Verizon IoT business. On the other hand, probably this figure makes Verizon one of the largest telecommunications operators worldwide in terms of IoT business in general and nonconnectivity IoT in particular. In any case, it represents only 0.5% of Verizon revenues ($150 million out of $30 billion in third quarter of 2014). Verizon product portfolio includes the following[100,101]:

- Connected car white label product for vehicle manufacturers (such as Mercedes Benz and Volkswagen), with customer value proposition similar to the ones from General Motors.

- Connected car aftermarket (i.e., not embedded in the car during manufacturing but as an external device acquired by end user afterward). There might be some limitations in terms of functionality (internal car systems accessible through external bus) and installation might be complex (a matter of tradeoff) and might need to be done in a car workshop. It includes a partnership with an auto insurer (StateFarm) offering discounts up to 50%.[102] Verizon Vehicle launching in September 2015, providing vehicle diagnostics and roadside assistance under a subscription business model. It requires to plug (apparently very easily) an on-board diagnostics reader (connected to On-Board Diagnostics OBD II port, compatible with most vehicles manufactured in the last 20 years) as well as a speaker (Bluetooth) for hands-free calling with buttons for emergency and help. User interface is complemented with optional smartphone application (available as usual for two main operating systems). Pricing is $14.99 per month, including equipment worth $120 (for a minimum contract of 2 years, or if cancelled earlier, equipment must be returned). Promotion full details can be seen on the product website.[103]

- Fleet management.

In 2014 Vodafone acquired Cobra Automotive Technologies for €145 million[104] (almost €200 million including debt). Based in Italy and with operations in Europe (six largest markets: France, Germany, Italy, Spain, the United Kingdom, and Switzerland), Asia (Japan, South Korea and China), and Latin America (Brazil), its customers are vehicle manufacturers, dealerships, and end customers (aftermarket) with services such as vehicle tracking, telematics, and usage-based insurance. Afterward in 2015, it was rebranded to Vodafone Automotive.[105] The two main business lines are telematics services and electronics systems (related to telematics, including antitheft and parking assistance). Further information can be found on its website.[106]

On a probably smaller scale, Orange, through its enterprise division (Business Services), acquired in 2015 a French company called Ocean, devoted to fleet management (45,000 from 2000 customers) and vehicle tracking solutions. Together, they manage over 100,000 vehicles. Ocean has been integrated as a unit of Orange Applications for Business (enterprise digital services division comprising M2M, big data analytics, and digitalization of customer experience).[107]

Another relevant telecommunications operator in the space that has already been mentioned is AT&T. As opposed to Verizon and Vodafone, it has not made a major acquisition. But as already stated in Section 6.1, AT&T got impressive, almost 700,000 connected cars, added to its roster in only one quarter (first quarter of 2015), representing more than half of wireless added in the same period, in addition to the 3 million that it already served in 2014.[108] Connected car efforts are part of AT&T IoT Solutions (formerly Emerging Devices) organization (more details on its website).[109] As already commented, the partnerships with car manufacturers is based on AT&T Drive platform, which is modular offering different features and capabilities based on the partner needs. The basic offering is connectivity, but others included billing, data analytics, infotainment content and applications, firmware upgrades over-the-air, and so on. Sample apps included in 2015 can be seen at this reference.[110] Besides these agreements with car vendors, AT&T also offers an aftermarket device, called Car Connection 2.0, that also connects to OBDII port, and is sold for $99.99 and requires connectivity contract (e.g., shared data plan) from AT&T.[111]

Sometimes a specific application is offered standalone, although the business case is more challenging, since the costs (specially the hardware that might be expensive) cannot be shared among different applications. In insurance telematics, there were around 11.6 million policies by the end of 2015, combining Europe (5.3 million, predominantly aftermarket hardwired blackboxes) and North America (6.3 million typically self-installed on-board-diagnostics devices).[112] As commented before, the growing number of connected cars would not require the owner to install additional devices, which could expand its adoption. As an alternative, smartphone-based applications could be a minimum viable product offering a low-cost alternative, although not as robust as a specific hardware embedded/connected to the car. Examples of leading niche providers of insurance telematics are Octo Telematics, Intelligent Mechatronic Systems, Baseline Telematics, Scope Technologies, Wunelli, Modus, MyDrive Solutions, DriveFactor, and so on.

The field of commercial telematics, which could be another way of calling fleet management (or more generally asset tracking), is probably the oldest application vertical, started even before the machine-to-machine term was coined. It was pioneered in the late 1980s by Qualcomm with its Omnitracs system. In 2013, Qualcomm sold its Omnitracs division with operations in the Americas (the United States, Canada, and Latin America) for $800 million in cash.[113] As example of pricing in the sector, Omnitracs has plans in the United States starting at $7.95 for the most basic, and at the other end, the premium one starts at $39.95, including features such as analytics, messaging, positions, hours of service, vehicle inspection, performance monitoring, critical event reporting, fault monitoring, driver workflow or trip manager, and so on.[114] On top of companies already mentioned, we could include Masternaut, Mix Telematics, Wireless Car, OrbComm, Trimble,

Telogis, Arvento, Navman Wireless, Ituran, Geotab, Sascar (acquired by tire manufacturer Michelin in 2014 for around €520 million including debt, with 33,000 fleets managed and 190,000 trucks in Brazil[115]), and so on. There are also companies traditionally associated with personal navigation devices such as TomTom. Intensive mergers and acquisitions are happening in this space, with room for consolidation. By the end of 2013 there were 3.65 million active fleet management systems in Europe, a fragmented market in which TomTom (400,000), Masternaut (350,000), and Digicore (100,000) are market leaders, together with vehicle manufacturers such as Daimler (150,000), Volvo (135,000), and Scania (100,000).[116] In Americas, there were almost 6 million (4 million in North America plus 1.9 million in Latin America) also by the end of 2013.[117] Again the market is quite fragmented, with hundreds of players despite ongoing consolidation, with top 10 providers accounting for less than half of market. Leaders with more than 300,000 units are Fleetmatics, Trimble, Zonar Systems, and Telogis. Still behind Americas and Europe, in China there were approximately 2.1 million units by the end of 2014,[118] typically with limited features. As usual the market is dominated by local players.

6.4.3 Platform

In this chapter, we have covered two typical industries (utilities with metering applications and automotive) for machine-to-machine applications. However, there is a very long tail of industries that can benefit from machine-to-machine connectivity and build on top of it with integrated applications. Industry behemoths can afford to build customized end-to-end applications, but in general companies can leverage horizontal application enablement platforms that offer generic modules common to the different industries (e.g., agriculture, manufacturing, mining, transportation, energy, and government), reducing time to market and providing access to existing applications offered by PTC together with its partners and ecosystem third party developers.

PTC is the major player in the space that has grown through acquisitions. In May 2015, it acquired the analytics firm ColdLight for around $105 million,[119] strengthening its position in the machine learning and predictive analytics space, one of the pillars of Internet of Things, leveraging the huge amounts of data generated by connected objects. In July 2014, it acquired Axeda for $170 million in cash,[120] with its technology for remotely managing sensors, machines, and devices securely, including over-the-air software upgrades. In December 2013, it acquired ThingWorx for approximately $112 million, a platform for rapid application development.[121] Telecommunications operators that have partnered with PTC include NTT Docomo in Japan,[122] Elisa in Finland and Estonia,[123] Telenor (through its Connexion subsidiary) in Central and Eastern Europe and Asia,[124] and so on. It also has partners in different industries such as health care (e.g., monitoring systems, personalized information), agriculture (e.g., livestock management), transportation (e.g., bike-share), manufacturing, and so on.[125] It also has educational partnerships, with over 200 universities and colleges in the world offering Internet of Things Academic Program that includes the application development platform hosted by PTC.[126]

6.5 ACRONYMS

2G/3G/4G/5G	2nd/3rd/4th/5th Generation
3GPP	3rd Generation Partnership Project
BoM	Bill of Materials
CAGR	Compound Annual Growth Rate
CDMA	Code Division Multiple Access
CIoT	Cellular Internet of Things
dB	decibel
dBm	decibel relative to milliwatt
EC-GSM	Extended Coverage GSM
EDGE	Enhanced Data rates for Global Evolution
EGPRS	Enhanced General Packet Radio Service
ETSI	European Telecommunications Standards Institute
eUICC	Embedded Universal Integrated Circuit Card
FDD	Frequency Division Duplexing
GPRS	General Packet Radio Service
GSM	Global System for Mobile
GSMA	GSM Association
HSPA	High-Speed Packet Access
IoT	Internet of Things
ISM	Industrial, Scientific and Medical
LAN	Local Area Network
LoRa	Long Range
LPWA	Low-Power Wide Area
LTE	Long-Term Evolution
M2M	Machine-to-Machine
mA/mW	milliampere/watt
MAC	Medium Access Control
MHz	megahertz
MTC	Machine-Type Communications
NB-IoT	Narrowband IoT
OBD	On-Board Diagnostics

OFDM	Orthogonal Frequency Division Multiplexing
RAN	Radio Access Network
RF	Radio Frequency
RPMA	Random Phase Multiple Access
SIM	Subscriber Identity Module
TR	Technical Report
UE	User Equipment
UMTS	Universal Mobile Telecommunications Service
UNB	Ultra-Narrowband
UTRAN	UMTS Terrestrial Radio Access NodeB
WAN	Wide Area Network

NOTES

1. Cellular M2M forecasts: unlocking growth. Silwia Kechiche, Lead Analyst M2M. GSMA Intelligence ©2015.https://gsmaintelligence.com/research/2015/02/cellular-m2m-forecasts-unlocking-growth/457/ (February 2015).
2. OnStar launches in Brazil on Chevrolet Cruze. General Motors, http://media.gm.com/media/us/en/gm/news.detail.html/content/Pages/news/us/en/2015/jun/innovation/connectivity/0624-onstar-brazil.html (June 24, 2015).
3. Telefónica UK signs £1.5 bn smart meter deal. O2 UK, http://news.o2.co.uk/?press-release=telefonica-uk-signs-1-5bn-smart-meter-deal (September 23, 2013).
4. World vehicle population tops 1 billion units. John Sousanis, WardsAuto © 2015 Penton,http://wardsauto.com/ar/world_vehicle_population_110815 (August 15, 2011)
5. AT&T reports industry-leading wireless postpaid churn, 1.2 million wireless net adds and strong demand for strategic business services in first-quarter results. AT&T, http://about.att.com/story/att_first_quarter_earnings_2015.html (April 22, 2015).
6. Bloomberg Business website. Bloomberg,//www.bloomberg.com/ (July 1, 2015), China Mobile //www.bloomberg.com/quote/CHL:US, Verizon //www.bloomberg.com/quote/VZ:US, AT&T //www.bloomberg.com/quote/T:US, Vodafone //www.bloomberg.com/quote/VOD:US, Nippon Telegraph and Telephone (NTT) //www.bloomberg.com/quote/NTT:US, Deutsche Telekom //www.bloomberg.com/quote/DTEGY:US, NTT Docomo //www.bloomberg.com/quote/DCM:US, America Movil //www.bloomberg.com/quote/AMX:US, Softbank //www.bloomberg.com/quote/SFTBF:US, Telefonica //www.bloomberg.com/quote/TEF:US, KDDI //www.bloomberg.com/quote/KDDIF:US, BT //www.bloomberg.com/quote/BT:US, Telstra //www.bloomberg.com/quote/TLSYY:US, Singapore Telecom (Singtel) //www.bloomberg.com/quote/SGAPY:US, China Telecom //www.bloomberg.com/quote/CHA:US, Orange //www.bloomberg.com/quote/ORAN:US, China Unicom //www.bloomberg.com/quote/ORAN:US, BCE //www.bloomberg.com/quote/BCE:US,
7. Global M2M market to grow to 27 billion devices, generating USD1.6 trillion revenue in 2024. Machina Research.https://machinaresearch.com/news/global-m2m-market-to-grow-to-27-billion-devices-generating-usd16-trillion-revenue-in-2024/ (June 24, 2015).
8. Acessos_SMP_2015–2016_-_Total.csv. AGÊNCIA NACIONAL DE TELECOMUNICAÇÕES (Anatel), http://ftp.anatel.gov.br/dados/Acessos/Movel_Pessoal/Total/csv/Acessos_SMP_2015-2016_-_Total.csv (July 17, 2015). (Brasil registra 281,70 milhões de linhas móveis em operação em janeiro de 2015. AGÊNCIA NACIONAL DE TELECOMUNICAÇÕES (Anatel), //www.anatel.

gov.br/Portal/exibirPortalNoticias.do?acao=carregaNoticia&codigo=36556March 6, 2015 (http://ftp.anatel.gov.br/dados/, http://ftp.anatel.gov.br/dados/Acessos/Movel_Pessoal/); Comunicação Máquina a Máquina (June 1, 2015). //www.anatel.gov.br/dados/index.php?option=com_content&view=article&id=282;Telefonia Móvel – Acessos, //www.anatel.gov.br/dados/index.php?option=com_content&view=article&id=270) (May 21, 2015).

9. Mobile Telephony, Lines, M2M lines (Monthly data February 2015). Comisión Nacional de Mercados y Competencias, http://data.cnmc.es:80/datagraph/jsp/informe_1_1.jsp?aqsdqsiid=14013&tipoinforme=6&periodicidad=3 (May 26, 2015).

10. Global price plans (Coverage and pricing). MyM2M. //www.mym2m.com/rates.php?lang=en&t=Global (last accessed July 5, 2015).

11. About Docomo>Company Information>Overview>History (*March 2012: Shifts customer base completely to 3G and LTE services*). NTT Docomo,//www.nttdocomo.co.jp/english/corporate/about/outline/history/ (last accessed July 5, 2015).

12. Discontinuation Notice of 2G Service. Softbank,//www.softbank.jp/en/corp/group/sbm/news/press/2009/20091124_01/ (November 24, 2009).

13. Last 2G phone shipped 8 months ago in Japan (*KDDI/AU switched off their 2G radio network in March this year*). Eurotechnology Japan KK ©2013,//www.eurotechnology.com/2008/08/22/last-2g-phone-shipped-8-months-ago-in-japan/ (August 22, 2008).

14. AT&T's Donovan: When it comes to 5G, timing is everything (Citi 2015 Global Internet, Media & Telecommunications Conference Tuesday in Las Vegas: *Donovan also talked briefly about the status of the company's 2G network, which it has said it will decommission in 2016. AT&T has been actively refarming the 2G spectrum for LTE and Donovan said that in most markets there is only 5 or 10 MHz of spectrum that is still devoted to the 2G network. The rest has already been refarmed*). FierceWireless, a publication of FierceMarkets, a division of Questex, LLC © 2015. //www.fiercewireless.com/tech/story/atts-donovan-when-it-comes-5g-timing-everything/2015-01-07 (January 7, 2015).

15. Telenor Norway—Network. Magnus Zetterberg, CTO, Telenor Norway. Telenor, //www.telenor.com/wp-content/uploads/2015/06/02-Telenor-Norway-Seminar-London-CTO-FINAL.pdf (June 2, 2015), p. 7.

16. The LTE revolution has only just begun and networks try to catch up (*John Horn, president of Cincinnati, Ohio-based Raco Wireless[. . .]baseline module cost is $10*). Canadian Business © 1999–2015 Rogers Media, //www.canadianbusiness.com/insights/the-lte-revolution-has-only-just-begun-and-networks-try-to-catch-up/ (September 25, 2014).

17. Despite North America, 2G M2M cellular module shipments remain strong as market neared $1.1 billion in 2013. ABI research, Allied Business Intelligence, Inc. ©2015, //www.abiresearch.com/press/despite-north-america-2g-m2m-cellular-module-shipm/ (July 23, 2014).

18. Sierra Wireless Q1 2015, ABI Research Report, July 2014. Sierra Wireless, //www.sierrawireless.com/~/media/pdf/investors/2015/sierrawireless_q1_2015_final.ashx?la=en (May 7, 2015) (//www.sierrawireless.com/aboutus/investorinformation/quarterlyresults/).

19. M2M cellular module vendors compete on volume as Sierra Wireless still leads the Pack in revenues. ABI research, Allied Business Intelligence, Inc., ©2015, //www.abiresearch.com/press/m2m-cellular-module-vendors-compete-on-volume-as-s/ (June 3, 2015).

20. M2M World Alliance-Members, M2M World Alliance, //www.m2mworldalliance.com/#section_members (last accessed July 5, 2015).

21. Leading M2M Alliances back the GSMA embedded SIM specification to accelerate the Internet of Things. GSMA, © GSM Association 1999 –2014,//www.gsma.com/connectedliving/news/leading-m2m-alliances-back-the-gsma-embedded-sim-specification-to-accelerate-the-internet-of-things/ (March 2, 2015).

22. Success of GSMA embedded SIM Spec in automotive sector will open up new markets. GSMA, © GSM Association 1999–2014, //www.gsma.com/connectedliving/news/success-of-gsma-embedded-sim-spec-in-automotive-sector-will-open-up-new-markets/ (June 12, 2015).

23. SGP.02 v3.0—Remote provisioning architecture for embedded UICC Technical Specification. GSMA, © GSM Association 1999–2014, //www.gsma.com/newsroom/all-documents/sgp-02-v3-0-remote-provisioning-architecture-for-embedded-uicc-technical-specification/ (June 30, 2015).

24. With 3 billion connections, LPWA will dominate wide-area wireless connectivity for M2M BY 2023, Machina Research, //machinaresearch.com/news/with-3-billion-connections-lpwa-will-dominate-wide-area-wireless-connectivity-for-m2m-by-2023/ (February 25, 2015).
25. Elster, //www.elstermetering.com/en/index (last accessed July 5, 2015).
26. Website. Amber-Wireless, //www.amber-wireless.com/en/portal.html (last accessed July 5, 2015).
27. Telensa, //www.telensa.com (last accessed July 5, 2015).
28. Senaptic turns IoT into real-world revenue streams, ©2015 Business Wire,, //www.businesswire.com/news/home/20140403005553/en/Senaptic-Turns-IoT-Real-World-Revenue-Streams#.VZEp1xvtlBc (April 3, 2015).
29. Telensa and Senaptic come together to build IoT future. Telensa, //www.telensa.com/newsarea/news/telensa-and-senaptic-come-together-to-build-iot-future (March 19, 2015).
30. M2M spectrum, //m2mspectrum.com/ (last accessed July 5, 2015).
31. SIGFOX, the global leader in Internet of Things connectivity, secures a record $115 million round of funding. ©2015 Business Wire, //www.businesswire.com/news/home/20150211005248/en/SIGFOX-Global-Leader-Internet-Connectivity-Secures-Record (February 11, 2015).
32. IoT start-up Sigfox launching 902 MHz network nationwide in U.S. FierceWireless a publication of FierceMarkets, a division of Questex, LLC © 2015, //www.fiercewireless.com/tech/story/iot-startup-sigfox-launching-902-mhz-network-nationwide-us/2015-03-03 (March 3, 2015).
33. ENGIE to Roll out SIGFOX Internet of Things Network in Belgium. ©2015 Business Wire, //www.businesswire.com/news/home/20150616005774/en/ENGIE-Roll-SIGFOX-Internet-Network-Belgium#.VYhK5_ntlBc (June 16, 2015).
34. SIGFOX adds Denmark to its global Internet of Things network. ©2015 Business Wire, //www.businesswire.com/news/home/20150609005767/en/SIGFOX-Adds-Denmark-Global-Internet-Network#.VYhK0PntlBc (June 9, 2015). (*This strategic investment, which includes leading mobile network operators, clearly demonstrates how SIGFOX's two-way low-throughput network complements existing high-bandwidth networks. The company sees a clear path toward unifying them in a single network, allowing always-efficient connectivity from both energy and throughput standpoints.*)
35. LTE for M2M and IoT—viable now with LTE Cat 1. Sequans Communications (April 3, 2015). (*Cat 0 devices may not be ready for mass deployment until 2017, and probably not until 2018 for Release 13 features.*)
36. Technology enablers. © 2014 SIGFOX, //www.sigfox.com/en/#!/connected-world/technology-enablers (last accessed July 5, 2015).
37. Technology. © 2014 SIGFOX, //www.sigfox.com/en/#!/technology (last accessed July 5, 2015).
38. Verizon Wireless, Ericsson and Sequans complete LTE category 1 device and network trial. Sequans Communications.//www.sequans.com/press-release/verizon-wireless-ericsson-sequans-complete-lte-category-1-device-network-trial/ (March 2, 2015).
39. Sequans completes trial of Cat 1 LTE technology with a second major US operator. Sequans Communications, //www.sequans.com/sequans-completes-trial-cat-1-lte-technology-second-major-us-operator/ (May 21, 2015).
40. LTE Cat M1/NB1 E on the horizon, while LTE Cat 1 is spreading around the globe. Sequans, //www.sequans.com/lte-cat-m1nb1-e-on-the-horizon-while-lte-cat-1-is-spreading-around-the-globe/ (June 30, 2016).
41. LTE for IOT has arrived: ALTAIR disrupts M2M market with new CAT-0, CAT-1 chipsets. © 2015 Altair Semiconductor, http://altair-semi.com/press/lte-for-iot-has-arrived-altair-disrupts-m2m-market-with-new-cat-0-cat-1-chipsets/ (February 25, 2015).
42. The wait is over: ALTAIR brings affordable, low power, robust LTE connectivity to IoT. By Paula Bernier, © 2015 IOT Magazine, //www.iotevolutionmagazine.com/features/articles/404696-wait-over-altair-brings-affordable-low-power-robust.htm (June 8, 2015).
43. Recent advancements in M2M communications in 4G networks and evolution toward 5G, Ratasuk, R. Nokia Networks, Arlington Heights, IL, Prasad, A.; Zexian Li; Ghosh, A.; Uusitalo, M. *2015 18th International Conference on Intelligence in Next Generation Networks* (ICIN), IEEE,, pp. 52–57. http://ieeexplore.ieee.org/xpl/login.jsp?tp=&arnumber=7073806 (February 17–19, 2015).

44. M2M—adapting LTE for the Internet of Things: Nokia LTE M2M optimizing LTE for the Internet of Things White Paper. Nokia, http://networks.nokia.com/news-events/insight-newsletter/articles/m2m-adapting-lte-for-the-internet-of-things (October 6, 2014).

45. Service requirements for Machine-Type Communications (MTC); Stage 1. © 2014, 3GPP Organizational Partners (ARIB, ATIS, CCSA, ETSI, TTA, TTC), //www.3gpp.org/DynaReport/22368.htm

46. Study on Machine-Type Communications (MTC) and other mobile data applications communications enhancements. © 2014, 3GPP Organizational Partners (ARIB, ATIS, CCSA, ETSI, TTA, TTC), //www.3gpp.org/DynaReport/23887.htm

47. Study on enhancements to Machine-Type Communications (MTC) and other mobile data applications; Radio Access Network (RAN) aspects. © 2014, 3GPP Organizational Partners (ARIB, ATIS, CCSA, ETSI, TTA, TTC), //www.3gpp.org/DynaReport/37869.htm.

48. Study on provision of low-cost Machine-Type Communications (MTC) user equipments (UEs) based on LTE. © 2014, 3GPP Organizational Partners (ARIB, ATIS, CCSA, ETSI, TTA, TTC), //www.3gpp.org/dynareport/36888.htm.

49. Evolved Universal Terrestrial Radio Access (E-UTRA); User equipment (UE) radio access capabilities. © 2014, 3GPP Organizational Partners (ARIB, ATIS, CCSA, ETSI, TTA, TTC), //www.3gpp.org/DynaReport/36306.htm.

50. LTE ue-Category. 3GPP, //www.3gpp.org/keywords-acronyms/1612-ue-category (last update July 10, 2014).

51. LTE UE category & class definitions. Radio-electronics.com, by Ian Poole. © Adrio Communications Ltd., //www.radio-electronics.com/info/cellulartelecomms/lte-long-term-evolution/ue-category-categories-classes.php (last accessed July 5, 2015).

52. © 2015 Weightless SIG, //www.weightless.org/ (last accessed July 5, 2015).

53. Weightless-N open standard goes live. © 2015 Weightless SIG, //www.weightless.org/news/weightlessn-open-standard-goes-live (May 5, 2015).

54. © 2015 Weightless SIG, //www.weightless.org/keyfeatures/5-km-range (last accessed July 5, 2015).

55. Spectrum flexibility. © 2015 Weightless SIG, //www.weightless.org/keyfeatures/spectrum-flexibility (last accessed July 5, 2015).

56. Weightless SIG shelves white spaces, but it has new Internet-of-Things spectrum in mind. GigaOm, Knowingly, Inc., https://gigaom.com/2014/09/17/weightless-sig-shelves-white-spaces-but-it-has-new-internet-of-things-spectrum-in-mind/ (September 17, 2014).

57. Hardware. © 2015 NWave, //www.nwave.io/ (last accessed January 7, 2015).

58. Hardware. © 2015 Weightless SIG, //www.weightless.org/about/hardware (last accessed July 5, 2015).

59. BT adopts weightless technology. © 2015 Weightless SIG, //www.weightless.org/news/bt-adopts-weightless-technology (January 15, 2014).

60. Huawei demonstrates pre-standard LTE-M, Vodafone targets Cellular IoT. © Copyright Rethink Research 2014,//www.rethinkresearch.biz/articles/huawei-demonstrates-pre-standard-lte-m-vodafone-targets-cellular-iot/ (March 6, 2015).

61. Vodafone extends its network capability to further support the Internet of Things. © 2015 Vodafone Group, //www.vodafone.com/content/index/about/what/technology-blog/2015/02/vodafone_extendsits.html (February 13, 2015).

62. GP-150083 - CIoT—Coexistence with GSM (update of GPC150020), //www.3gpp.org/ftp/tsg_geran/TSG_GERAN/GERAN_65_Shanghai/Docs/GP-150083.zip (March 2015).

63. © 2015 LoRa Alliance,http://lora-alliance.org/ (last accessed July 5, 2015).

64. LoRaWAN R1.0 open standard released for the IoT. © 2015 LoRa Alliance. http://lora-alliance.org/kbdetail/Contenttype/ArticleDet/moduleId/583/Aid/23/PR/PR (June 16, 2015).

65. Internet of Things specialist Actility announces $25 million funding round led by Ginko Ventures, KPN, Orange, Swisscom and Foxconn. © Actility 2014, //www.thingpark.com/en/news/internet-things-specialist-actility-announces-25-million-funding-round-led-ginko-ventures-kpn-0 (June 16, 2015).

66. The rise of the machines—world's first global standards for M2M deployment. © Copyright 2015 oneM2M partners, //www.onem2m.org/news-events/news/53-the-rise-of-the-machines-world-s-first-global-standards-for-m2m-deployment (February 4, 2015).

67. Bouygues Telecom announces June launch of France's first "Internet-of-Things" network based on LoRa technology. © 2015 LoRa Alliance, //lora-alliance.org/kbdetail/Contenttype/ArticleDet/moduleId/583/Aid/20/PR/PR (March 26, 2015).
68. 802.11ah: improving whole home coverage and power efficiency. ©2015 Qualcomm Technologies, Inc., //www.qualcomm.com/invention/research/projects/wi-fi-evolution/80211ah (last accessed July 5, 2015).
69. Official IEEE 802.11 working group project Timelines. Copyright Institute of Electrical and Electronics Engineers, Inc., //www.ieee802.org/11/Reports/802.11_Timelines.htm (last accessed July 5, 2015).
70. IEEE P802.11—Task Group AH—meeting update. Status of project IEEE 802.11ah. Copyright Institute of Electrical and Electronics Engineers, Inc., //www.ieee802.org/11/Reports/tgah_update. htm (last accessed July 5, 2015).
71. IEEE P802.11ah/D5.0, March 2015—draft. Copyright 2013, IEEE-SA, //www.techstreet.com/ieee/products/1880837#jumps (March 2015).
72. IEEE 802.11AH: the WiFi approach for M2M communications. Adame, T. Univ. Pompeu Fabra, Barcelona, Spain; Bel, A.; Bellalta, B.; Barcelo, J.; Oliver, M. IEEE *Wireless Communications Magazine*, December 2014, http://ieeexplore.ieee.org/xpl/articleDetails.jsp?arnumber=7000982.
73. Outdoor long-range WLANs: a lesson for IEEE 802.11ah. Aust, S. NEC Communication Systems, Ltd., Kawasaki, Japan; Prasad, R.; Niemegeers, I.G.M.M. IEEE Communications Surveys and Tutorials, May 2015, http://ieeexplore.ieee.org/xpl/articleDetails.jsp?arnumber=7101216.
74. On-ramp wireless expands Internet of Things (IoT) connectivity with over 30 global network deployments. © Copyright 2015, On-Ramp Wireless, Inc., //www.onrampwireless.com/on-ramp-wireless-expands-internet-of-things-iot-connectivity-with-over-30-global-network-deployments/ (April 14, 2015).
75. The global IoT market opportunity will reach US$4.3 trillion by 2024. Machina Research, https://machinaresearch.com/news/the-global-iot-market-opportunity-will-reach-usd43-trillion-by-2024/ (April 21, 2015).
76. IBM connects "Internet of Things" to the enterprise. IBM, //www-03.ibm.com/press/us/en/pressrelease/46453.wss (March 31, 2015).
77. Intel reports record full-year revenue of $55.9 billion. Intel, //www.intc.com/releasedetail.cfm?ReleaseID=906520&ReleasesType= (January 15, 2015).
78. Qualcomm says Internet of Things effort already paying off. Bloomberg Business,//www.bloomberg.com/news/articles/2015-05-14/qualcomm-says-internet-of-things-effort-already-paying-off (May 14, 2015).
79. Second quarter Fiscal 2015 earnings. *Business Outlook*, Current Guidance FY 2015 estimates $25–7B. Qualcomm, http://files.shareholder.com/downloads/QCOM/441121093x0x822989/EAD6A29F-6244-486A-B5FB-15972BA64899/Q2FY15_Earnings_Executive_Presentation.pdf (April 22, 2015).
80. Qualcomm expands industry collaboration to grow the Internet of everything. Qualcomm, https://www.qualcomm.com/news/releases/2015/05/14-0 (May 14, 2015).
81. GE to open up Predix industrial Internet platform to all users. General Electric, //www.genewsroom.com/press-releases/ge-open-predix-industrial-internet-platform-all-users-278755 (October 9, 2015).
82. Services & Industrial Internet Investor Meeting. General Electric, //www.ge.com/investor-relations/ir-events/services-industrial-internet-investor-meeting (October 9, 2014).
83. GE sees fourfold rise in sales from Industrial Internet. Bloomberg Business, //www.bloomberg.com/news/articles/2014-10-09/ge-sees-1-billion-in-sales-from-industrial-internet (October 9, 2014).
84. Website. MirrorLink,. //www.mirrorlink.com/ (last accessed July 2, 2015).
85. Apple-CarPlay. Apple, //www.apple.com/ios/carplay/ (last accessed July 2, 2015).
86. Android Auto. Google, //www.android.com/auto/ (last accessed July 2, 2015).
87. Apple rolls out CarPlay giving drivers a smarter, safer & more fun way to use iPhone in the car. Apple, //www.apple.com/pr/library/2014/03/03Apple-Rolls-Out-CarPlay-Giving-Drivers-a-Smarter-Safer-More-Fun-Way-to-Use-iPhone-in-the-Car.html (March 3, 2014).

88. Android and the Open Automotive Alliance shift into the next gear. Open Automotive Alliance, //www.openautoalliance.net/#press (June 25, 2015).
89. OnStar trial. General Motors, https://www.onstar.com/us/en/support/helptopics.html#trial duration (last accessed July 3, 2015).
90. Insurance discounts. General Motors, https://www.onstar.com/us/en/services/insurance-discounts.html (last accessed July 3, 2015).
91. OnStar to offer driving feedback; customers can seek insurance discounts. General Motors, http://media.gm.com/media/us/en/gm/news.detail.html/content/Pages/news/us/en/2015/Jan/0104-smart-driver-gm.html (last accessed April 1, 2015).
92. Website OnStar. General Motors, https://www.onstar.com (last accessed July 2, 2015).
93. OnStar to offer services in Mexico in 2013. General Motors, http://media.gm.com/media/us/en/gm/news.detail.html/content/Pages/news/us/en/2012/Sep/0906_onstar_mx.html (September 6, 2012).
94. Europe premiere at the Geneva motor show: Opel OnStar to set new standards in vehicle safety and connectivity. Opel-General Motors, http://media.gm.com/media/intl/en/opel/news.detail.html/content/Pages/news/intl/en/2015/opel/03-04-onstar.html (April 3, 2015).
95. Opel OnStar. Your personal connectivity and service assistant. Opel-General Motors, //www.opel.com/onstar/onstar.html (last accessed July 3, 2015).
96. OnStar chooses network provider for global expansion. General Motors, http://media.gm.com/media/us/en/onstar/news.detail.html/content/Pages/news/us/en/2012/Feb/0223_telefonica.html (February 23, 2015).
97. GM's 4G connection to boost consumer features, revenue. Melissa Burden and Henry Payne, *The Detroit News*, June 18, 2015. //www.detroitnews.com/story/business/autos/general-motors/2015/06/18/gms-connection-boost-consumer-features-revenue/28964127/
98. OnStar For My Vehicle. General Motors, https://www2.onstar.com/web/fmv/home?g=1 (last accessed July 12, 2015).
99. Verizon, //www.verizonwireless.com/news/article/2012/06/pr2012-06-01.html (May 31, 2012).
100. Verizon connected car business looks for growth through platforms. *Wall Street Journal*, May 7, 2015, http://blogs.wsj.com/cio/2015/05/07/verizon-connected-car-business-looks-for-growth-through-platforms/
101. Drive connected. Verizon Telematics—Verizon Enterprise Solutions, https://www.verizontelematics.com/ (last accessed February 1, 2015).
102. In-drive. Verizon Telematics, //www.in-drive.com/sf/#IL (last accessed July 3, 2015).
103. Innovative roadside assistance | Verizon Vehicle. Verizon, https://www.verizonvehicle.com/ (last accessed July 5, 2015).
104. Vodafone intends to launch a voluntary takeover offer for Cobra Automotive Technologies S.p.A (Vodafone to complete acquisition of Cobra Automotive Technologies S.p.A. Vodafone), June 16, 2014, //www.vodafone.com/content/index/media/vodafone-group-releases/2014/cobra-offer.html (//www.vodafone.com/content/index/media/vodafone-group-releases/2014/vodafone_complete_acquisition_cobra.html) (August 7, 2015).
105. Introducing Vodafone Automotive: the new name for Cobra Automotive Technologies. Vodafone, http://m2m.vodafone.com/cs/m2m/insight_news/vodafone-automotive-the-new-name-for-cobra-automotive-technologies-2015-04-01 (April 1, 2015).
106. Vodafone Automotive, Delivering connected car services to the automotive and insurance industries?. Vodafone, http://automotive.vodafone.com/ (last accessed July 5, 2015).
107. Orange Business Services acquires Ocean to strengthen its vehicle fleet management activities. //www.orange.com/en/press/Press-releases/press-releases-2015/Orange-Business-Services-acquires-Ocean-to-strengthen-its-vehicle-fleet-management-activities
108. 2014 Annual Report: Connected Car. AT&T, //www.att.com/Investor/ATT_Annual/2014/connected_car.html (February 10, 2015).
109. AT&T Internet of Things. AT&T, //www.att.com/edo/ (last accessed July 5, 2015).
110. AT&T introduces three new applications to the AT&T Drive Connected Car platform,T&T Consumer Blog. AT&T, http://blogs.att.net/consumerblog/story/a7797754(March 2, 2015).
111. Connected Car from AT&T. AT&T, //www.att.com/shop/wireless/connected-car.html (last accessed July 5, 2015).

112. Europe and North America reached 11.6 million insurance telematics policies at year-end 2015. Berg Insight, //www.berginsight.com/News.aspx?m_m=6&s_m=1 (June 17, 2016).

113. Qualcomm finalizes sale of Omnitracs, Inc. to Vista equity partners. Qualcomm, https://www.qualcomm.com/news/releases/2013/11/25/qualcomm-finalizes-sale-omnitracs-inc-vista-equity-partners (November 25, 2013).

114. Omnitracs' application bundle plans matrix. Omnitracs, http://fleet.omnitracs.com/rs/omnitracsllc1/images/LCL1271-PRI_04-14_ApplicationBundlePlans_Matrix.pdf (last accessed July 13, 2015).

115. Michelin acquires Sascar, Brazil's leading digital fleet management company. Michelin, //www.michelin.com/eng/media-room/press-and-news/press-releases/Finance/Michelin-acquires-Sascar-Brazil-s-leading-digital-fleet-management-company (June 9, 2014).

116. The installed base of fleet management systems will reach 7.1 million in Europe by 2018. Berg Insight, //www.berginsight.com/News.aspx?m_m=6&s_m=1 (October 14, 2014).

117. The installed base of fleet management systems will reach 12 million units in the Americas by 2018. Berg Insight, //www.berginsight.com/News.aspx?m_m=6&s_m=1 (October 2, 2014).

118. The installed base of fleet management systems in China will reach 5.9 million units by 2019. Berg Insight, //www.berginsight.com/News.aspx?m_m=6&s_m=1 (January 27, 2015).

119. PTC to acquire big data machine learning and predictive analytics leader ColdLight.. //investor.ptc.com/releasedetail.cfm?ReleaseID=910818 (May 5, 2015).

120. PTC to acquire Axeda to expand Internet of Things technology portfolio. PTC, http://investor.ptc.com/releasedetail.cfm?ReleaseID=861634 (July 23, 2014).

121. PTC acquires leading Internet of Things platform provider ThingWorx. PTC, http://investor.ptc.com/releasedetail.cfm?ReleaseID=816468 (July 23, 2014).

122. PTC and NTT DOCOMO to extend uses of ThingWorx IoT development solution. PTC, http://investor.ptc.com/releasedetail.cfm?ReleaseID=911218 (May 6, 2015).

123. Elisa launches new Internet of Things (IoT) service with ThingWorx® Rapid Application Development Platform. PTC, http://investor.ptc.com/releasedetail.cfm?ReleaseID=891831 (January 19, 2015).

124. ThingWorx and Telenor Connexion launch collaboration to accelerate introduction of Innovative Connected Business Solutions. PTC, http://investor.ptc.com/releasedetail.cfm?ReleaseID=854765 (June 16, 2014).

125. News releases. PTC, http://investor.ptc.com/releases.cfm (last accessed July 13, 2015).

126. Over 200 global universities and colleges now offer the PTC Internet of Things academic program. PTC, http://investor.ptc.com/releasedetail.cfm?ReleaseID=912869 (May 13, 2015).

Chapter 7

IT: Cloud

Stefan Wesner

The role of IT (Information Technology) as a key element for delivering not only feature-rich but also cost-effective service offerings has increased over the last decade significantly. With emerging trends such as IoT (Internet of Things) or connected systems, for example, in manufacturing with Industry 4.0, not only the importance but also the need for flexibility will further increase.

Due to the important role of IT in delivering a service and the corresponding high investment costs, it is still common to design the service offer based on the limitations of the available IT infrastructure capabilities. This approach is obviously limiting the possibilities for a service portfolio however. Following the concept expressed by Richard Hunter from Gartner,[1] *the right way to discuss IT perform-ance and value is to focus on IT's contributions to business performance and busi-ness outcomes, and not on the performance of IT's machines,* a valuable IT infrastructure must follow business needs and be aligned with its goals. As business goals and priorities change over time, this does not only mean that an IT infra-structure must be designed to meet business goals but also adapt to change in a flexible and fast way to changing demands. Consequently, the process of business–IT alignment is established for many years in a range of industrial areas.

Traditional approaches to build customized IT infrastructures over a physical infrastructure with servers, networks, and corresponding software stack need weeks if not months. This is no longer meeting the needs of the fast changing business demands, customer diversity, and increasingly shorter time to market. The emer-gence of virtualization technology on desktops introduced by VMware in the late 1990s has found its way into the data center in a wide range of different flavors. Alone with virtualization a significantly increased IT infrastructure flexibility can be reached as it decouples the logical server infrastructure from the underpinning physical infrastructure. However, for fully utilizing the capabilities of a virtualized infrastructure, the concept of IaaS (Infrastructure as a Service) has emerged as an initial characteristic of what is commonly referred to as Cloud Computing nowa-days. However, the development has not stopped on the infrastructure level and a whole Cloud Stack build from layers is now widely accepted. Following the

Digital Services in the 21st Century: A Strategic and Business Perspective, First Edition.
Antonio Sánchez and Belén Carro.

definition from NIST,[2] the Cloud Computing Stack has been extended with a PaaS (Platform as a Service) offering on top of the IaaS layer basic services to be integrated into Cloud enabled application and the SaaS (Software as a Service) layer delivering the applications on top of a flexible IT infrastructure toward the customer. Beyond these well-established terms following the discussion above, the term BPaaS (Business Process as a Service)[3] had been introduced expressing the idea that a whole business process is delivered on top of one or more Cloud infrastructures utilizing the flexibility and capabilities to react to the changing IT demands of the underpinning Cloud services. Furthermore, the term NaaS (Network as a Service)[4] has been introduced to cover the functionality of virtualized network services within the Data Center, between several Data Centers and between Customer and the Data Center ensuring certain network capabilities from performance up to nonfunctional requirements such as isolated communication, for example, based on VXLAN tunnels. Often the NaaS functionality is realized relying on an SDN (Software Defined Networking)-capable network infrastructure.

The following sections are organized as follows: Starting from a deeper analysis of the global trends driving the evolution of Cloud computing and its basic delivery models are introduced and compared. The following section covers the virtualization concept as the key enabling technology preparing the more detailed discussion of the NIST-based layered Cloud model.

Based on this baseline, additional emerging approaches, in particular BPaaS, NaaS, and how Big Data Analytics as an important Cloud application impacts the further evolution of Cloud infrastructures are discussed in the following sections.

The chapter concludes with a discussion on emerging trends on data centre hardware and how this applies to the discussed concepts in this chapter. The chapter also covers some European research projects working on future Cloud architectures.

7.1   GLOBAL TRENDS DRIVING THE CLOUD EVOLUTION

As outlined earlier, the quality of the IT (Information Technology) infrastructure has visible impact on the quality of the delivered services. The traditional categories such as hardware mainly described by its computing and storage capacity, the software stack, and a set of fixed or slowly changing list of professional services on top of it are not sufficient to describe the flexibility expected nowadays.

A major disruption to IT has been Cloud which progressively is replacing legacy IT. Cloud is based on a per use business model, replacing the initial capital expenses by operating expenses (pay what you use) for a customer who no longer needs to invest in hardware anymore. For the telecommunication sector the offered services do not consist of monolithic offerings but a larger number of dependent services combined with communication services. Additionally, functional requirements for telcos (telecom operators) include functions such as security, especially related to networks and the combination of them with the compute and storage services.

7.1.1 Cloud Deployment Models

In the NIST report[5] three Cloud deployment models are introduced. The first model is the *Public Cloud*, an infrastructure hosted by an organization that is open to public use and used simultaneously by potentially many customers. The term *Private Cloud* refers to a model where a Cloud-based infrastructure is operated for a single organization. Besides these basic models, the concepts *Hybrid Cloud* and *Community Cloud* are introduced. A Community Cloud is understood as an extension of a private Cloud where consumers and/or providers share a common concern and operate a Cloud infrastructure jointly for their own use. The resources and consumers can be geographically dispersed. The Hybrid Cloud refers to composition of different Cloud Infrastructures of the categories outlined earlier. In such a scenario, parts of the services might be operated solely on the Private Cloud, whereas others can be migrated to a Public Cloud to cope with bursts in workloads, and address robustness and fault tolerance concerns or performance optimization such as load balancing.

These models are not meant to be taken as blueprints for deploying a Cloud infrastructure but as coarse-grained models to be instantiated for a specific use case in practice. A number of more fine-grained and elaborated models to differentiate the offers exist such as discussed in Endnote 6.[6]

7.1.2 Cloud Adoption

IT players (such as IBM, Microsoft, and SAP) are transforming themselves in migrating to Cloud. But new entrants based purely on Cloud are emerging strongly, for example, Amazon in IaaS and Salesforce in SaaS (mainly in specific segment of Customer Relationship Management). IaaS requires very heavy upfront investments in data center. Some telcos (such as NTT or Verizon) are trying to compete through acquisitions. For SaaS, telcos try to offer through a set of curated application, especially for SMEs. As the infrastructure or "Hardware as a Service" is well developed and does not allow easily to differentiate from the competition, recently the focus is more on higher layers either by providing building blocks for a problem domain such as big data analytics or machine learning capabilities as a service (e.g., Amazon Machine Learning launched in April 2015).[7]

7.2 VIRTUALIZATION AS ENABLING TECHNOLOGY

The concept of abstracting the physical hardware with an intermediate layer, providing to an operating system a virtualized computer, has been defined in the early 70s by Goldberg.[8] In this thesis, two fundamental models of hypervisors are defined. The bare-metal hypervisor (often referred to as type-1 hypervisor) is an intermediate layer between the physical hardware and potentially

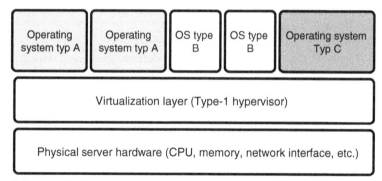

Figure 7.1   Bare-metal virtualization model.

many operating systems. The hosted hypervisor (also named type-2 hypervisor) is running on top of a host operating system. While the hosted approach is common for desktops where apart from a main operating system an additional environment is needed (e.g., for safe browsing or an OS of a different type), the bare-metal approach, as shown in Figure 7.1, is the common way to implement virtualization on server platforms. While the introduction of an abstraction layer between physical server and operating systems has obvious benefits in terms of flexibility and lifecycle of IT infrastructures, the performance penalties had been a concern.

However, with the support for virtualization in CPU design and similarly with virtualization support in other hardware devices, this overhead is reduced to a low overhead of a couple of percent that is acceptable for most applications, in particular comparing it with the huge benefits in flexibility.

7.3   THE LAYERED CLOUD MODEL

7.3.1   IAAS (Infrastructure as a Service)

In the previous section, a couple of challenges have been discussed with virtualized server infrastructures. The IaaS operates typically on top of a virtualized server infrastructure of several server systems and offers as additional key properties: *elasticity* as reaction on changing resource demands, *availability* upon observed hardware errors, and failures as well as an *optimized utilization* of the resources.

Elasticity is understood as the capability to dynamically allocate additional resources or reduce them as needed. This means that a predefined virtual server instance is created on one of the physical servers, configured automatically (e.g., setting up the network properties such as local IP addresses and virtual LAN ids) based on a predefined system image or as a clone of a running virtual server instance. It is important to understand that elasticity is different from scalability. Additional virtual servers and with them additional resources do *not* automatically lead to an increased ability to serve more requests or support more customers (Figure 7.2).

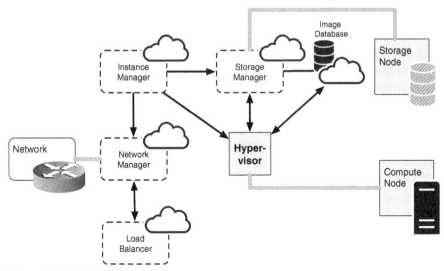

Figure 7.2 IaaS base architecture.

This means that an application must be capable of performing actions such as workload redistribution between the new number of available instances and also consolidating workload in case of reduced instances. Furthermore, elasticity can deliver an added value only if there is need to provide more resources (or destroy them if no longer needed) and if sufficient data on resource utilization can be collected and interpreted to trigger the corresponding actions. Figure 7.3 shows an example of workflow where a request to the services provided in virtual machine A is received. If no VM has been deployed yet it is created (decision point 1), system utilization (CPU, memory, network) is compared with predefined boundaries (decision point 2) and then either forwarded to an existing instance or the creation of a new one is triggered.

Based on this rule-based adaptation, the availability of services can be improved. Similarly, failure of a virtual system would be treated in the simplified example above in the same way as an overloaded or nonexistent instance. The remaining challenge is to find a good or even best matching physical host for the placement of a new instance of a virtual system. Also during operation due to workload changes the virtual system might need to be migrated to different physical hosts or new hosts must be started to cope with an increased workload. In many open source or commercial systems, the rules and deployment optimization approaches are rather simple and are still a research issue.[9,10] Another important aspect is that the whole control process must consider that the creation of additional instances might take significant time. Considering a base image of a full server image for a database, image might be several gigabytes in size. The image must be transferred to the selected physical host and started. Next step is that a data redistribution and/or replication is necessary. This does not easily cover minutes until an improved performance for accessing the

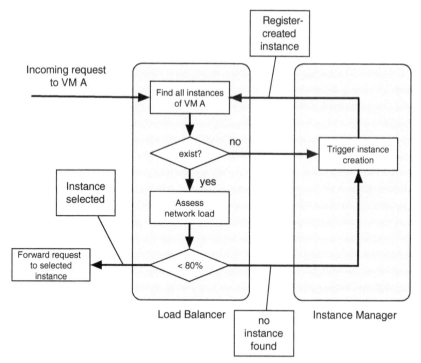

Figure 7.3   Example for rule-based adaptation to changing workload.

clustered database can be realized, but it has a negative impact on the performance during the migration process (Table 7.1).

7.3.2 PAAS

One of the reasons for the fast adoption of the IaaS concept is the low entry barrier
migrating from a physical server infrastructure. Applications running on top of an
IaaS infrastructure must not be changed and can be completely Cloud agnostic. The
IaaS offers still provide the notion of a virtual server as a single more or less independent instance from other virtual servers. While the Cloud infrastructure offers features
such as grouping a couple of instances, for example, for a given tenant or application
type, and allows the creation of the instance from a set of predefined instances as
"clones," the different instances are rather independent. As already outlined, capabilities such as workload redistribution or coping with increasing or decreasing number
of instances are left to the application logic. On such a virtual instance, potentially a
large number of different applications can be hosted running at the same time in con-
currence with the provided resources of a virtual server instance.

For a PaaS environment the notion of a server is further abstracted by providing an environment for executing an application. So the deployment process on
PaaS requires the user to package the application in a form that fits to the execution

Table 7.1 Key Components of an IaaS Infrastructure

Component	Responsibilities
Image Database	The Image Database contains a set of predefined images that can be configured and instantiated on demand by the Instance Manager
Instance Manager	The Instance Manager is responsible for creating new instances by getting the Image from the Image Database via the Storage Manager and deploy it on the most appropriate compute server
Storage Manager	The Storage Manager builds the interface between the Cloud system images and the Instance Manager. This is also necessary as not all images might be available for all customers and physical storage of Images might be replicated to different locations
Network Manager	The Network Manager is responsible for setting up the necessary elements to ensure that communication across VMs of one customer is isolated from another, for example, by using tunneling protocols such as VLAN or VXLAN as well as that internally used addresses are mapped to externally reachable addresses for the VMs that should be reachable from the outside
Load Balancer	As shown in Figure 7.3, the observed workload is continuously monitored and a rule-based decision is taken if new instances are necessary, where they have to be placed or if the number of instances can be reduced as workload has been decreasing

environment of the provider and a set of constraints/rules describing workload characteristics and how the provider is expected to react on changes of the workload. How the provider is implementing the execution environment for the application, how the scalability is achieved, and how the given constraints are considered is not transparent for the user. For example, if the PaaS provider delivers an environment for running very large Java applications by running virtual machines on very large shared memory systems or as a distributed virtual machine across several servers with a virtual shared memory, it is completely hidden from the PaaS user.

Besides single application instances, one can also deploy several applications communicating over provided channels such as message queues and allow the PaaS provider to optimize the actual deployment for a group of deployed application "cloudlets." While the abstraction level of IaaS is very similar to the physical host, the vendor lock-in is rather unlikely as despite differences in specific elements, such as a scalability rule language, the offers are conceptually very similar. The PaaS market is far more diverse and is offering solutions not only for potentially large number of application types but also sector or application specific solutions. A complex application consisting of several application elements and corresponding communication between them deployed, for example, to Google App Engine, is much harder to migrate to RedHat's OpenShift or to other general-purpose platforms such as Cloud Foundry as the migration steps necessary to move from IaaS Provider Amazon EC2 to an OpenStack environment as the basic abstraction level of the virtual server is well understood and defined, whereas the abstraction level of PaaS is more vendor dependent and potentially even tailored to specific application domain. An example for such a domain-specific environment relevant for telcos is a platform hosting NFV (Network Function Virtualization) services. As an example, a deep packet inspection service hosted on the platform would be automatically scaling up and down with increasing traffic or will be dependent on the amount of suspicious packets and the corresponding inspection level.

7.3.3 SAAS

The next level of abstraction proposed in the NIST architecture is the provision of full software services on demand. The SaaS (Software as a Service) model looks at initial glance conceptually very similar to the ASP (Application Service Provision) model. A more detailed look, however, clearly reveals that in contrast to the ASP model, where applications designed for a local server environment are provided by an external provider, the SaaS model goes far beyond that by realizing an environment designed from scratch to be executed remotely, used by multiple tenants not only separated but also collaboratively (e.g., such as collaborative editing on Google Docs or Microsoft's Office 365). In summary, the key difference is the business model for providing the application services. The ASP model assumes a per customer environment and consequently increases visibly the effort on the provider side for every new customer and makes also a fair sharing of the provided physical resources complex and cost-intensive. The SaaS model relies on the underpinning Cloud features such as elasticity, multitenancy, and decoupling from the physical hardware to allow a far easier scalability. So the core conceptual difference is that there is logically *one* software system for *all* customers (or at least customer segments) rather than copies/clones for different customers. For example, sharing a document in Google docs is not limited but can be done ad hoc with everybody that has an e-mail address.

While in the IaaS model the delivered service ends with the delivered hardware or a bare operating system, the platform model delivers execution environments for applications or application components targeting system administrators or application developers, and the SaaS model directly targets the end user.

7.3.4 An Illustrative Example

Realizing a blog space for a customer using the IaaS model would mean selecting a base image with a blog-hosting software, maybe an additional one with a fitting database installed and so on or selecting a base OS image and installing the package ourselves. Later we would add additional components or add software packages to our images to deal with DDoS, misuse of registration, commenting, and so on. We would need to think about instance types, sizes, and potential scalability rules translating increasing requests in changing hardware demands. In the PaaS model we would have blog space building blocks, such as the web-frontend, the persistency services, potentially other services dealing with authentication/authorization, DDoS prevention, registration, and would create our blog space from these blocks. Our PaaS would offer us to defined scalability rules on a more abstract, HW agnostic manner more driven by KPIs of our application. In this situation we would still have a system built just for us and our customers/users of the blog space.

The SaaS model would foresee a blog space provider system where we could get a certain service quality (e.g., space, requests, tools supporting the design/implementation). The space would allow us to open access to parts of the space to other users of the same service, potentially offers elements to aggregate/consolidate or index our content together with others and provide us a rich set of statistics analyzing visibility and allow us to compare this anonymously with statistics of other users.

7.4 ADVANCED CLOUD MODELS

The widely adopted classification splitting Cloud services into infrastructure, platform, or software provision as outlined earlier is only partially covering the wide variety of Cloud service models and many offers also do not clearly match with the categories but are a mixture of them.

7.4.1 Business Processes as a Service

As of today, the IaaS model has found wide adoption as alternative to traditional server hosting no matter if positioned as alternative to externally hosted physical or virtualized servers or as a private Cloud for internal users. Similarly, the SaaS model delivering a ready to use solution has been adopted in a couple of application domains. The PaaS positioned in the middle as outlined earlier delivering building blocks to design tailored and specific applications or platforms is still lacking behind in its adoption. While lots of simplifications had been integrated in the corresponding platforms, for example, to run Java-based applications, it still addresses the technology-oriented customer's ability to build its applications based on the platform services. As a result, there is a gap between the ready designed and only little adaptable SaaS model and the fully flexible, but technology-oriented IaaS and PaaS approaches (Figure 7.4).

The notion of Business Process as a Service (BPaaS) aims to position itself in between the fully flexible IaaS/PaaS solutions and the rather fixed SaaS models. The driving idea is that small- and medium-sized enterprises, in particular, do not have the human resource capacity to cope with the complexity of Clouds and rely on the services of a Cloud Broker that is delivering certain business level functionality expressed, for example, in a BMPN 2.0 model and amended with quality criteria or Key Performance Indicators (KPI) on business level and realize them using Cloud services.

As of today, the huge semantic gap between the business process and its corresponding business level constraints and the technical infrastructure requires

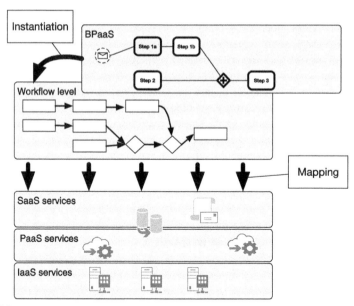

Figure 7.4   BPaaS principle.

significant effort of the Cloud Broker. The selection of appropriate Cloud services from IaaS, PaaS, and SaaS providers needed to realize and implement the process and orchestrate them into an efficient workflow is a complex task that requires in-depth knowledge of the Cloud Broker about the offered components and services on the platforms.

7.4.2   Network as a Service

The performance overhead by using virtual cores/CPUs has decreased tremendously and can now be neglected for most applications. Network virtualization in default Cloud applications such as vanilla Open Stack rely on software switch solutions such as Open vSwitch, emulating a switch with ports dedicated to virtual machines and acting as intermediary to the physical network interface cards. While this approach delivers very high flexibility, for example, for migrating VMs across servers and with them all their network configuration, the emulation comes at the expense of limited bandwidth and increased latency. If the storage attached to virtual machines is not realized with local storage devices but as a shared file system connected over the network, these limitations quickly become a bottleneck.

This problem is addressed from two angles. First, the communication affinity between VMs are considered, while making a placement decision such as increased connectivity aims to place them on servers close to each other or ideally even on the same host. An alternative approach is to realize the shared network connectivity relying on the Single Root I/O Virtualization (SR-IOV). Using these techniques, a single PCIe device can be virtualized using features of the network interface card to appear as several PCIe devices. Using this functionality, each Virtual Machine can be attached to such a virtual channel/device without the need to have a software-based emulation and the expected corresponding increased performance similar to a nonvirtualized environment. However, using this approach, the mobility of a VM is limited and a transfer of a VM between servers is not easily possible anymore and is therefore only recommended for environments where VM reallocation happens frequently.

Network as a Service functionality is not limited to a Data Center environment. In particular in scenarios where Cloud services are delivered by several Cloud service providers, the communications between the different providers need special attention if certain performance capabilities are necessary. Besides the regular layer 3/4 connectivity between two Cloud providers or dedicated fiber connectivity, a couple of tunneling protocols such as VXLAN or VPN over GRE are common solutions to implement virtual layer 2 connectivity. Such functionality needs extension of typical Cloud middleware components such as the Neutron module of OpenStack in order to implement such functionality within an Internet Service Provider (ISP) domain or even between ISPs. A couple of different solution approaches are in use. Beyond manual setup in particular, Virtual Private LAN Services (VPLS)- and Software Defined Networking (SDN)-based approaches are in use to ensure that the Data Center to Data Center connectivity is on the level needed for an efficient distributed Cloud application.

7.5 FUTURE CLOUD MODELS

7.5.1 Increased Variety and Complexity of Data Center Infrastructure

In the near future, the variety and complexity of available hardware for building a data center, and with them future Cloud service providers, will increase significantly. This trend that has started with the move toward multicore CPUs in 2003, driven by the need to increase performance but at the same time deliver better energy efficiency, is currently evolving as follows:

- As the increased number of cores has increased the floating point operations per second within a CPU much faster as the interface to the main memory of the server systems (also known as the "Memory Wall"), the boundary between storage and memory will disappear and a more complex memory hierarchy will evolve. This deep memory hierarchy from Cache over Main Memory of different flavors (High Bandwidth Memory, Capacity Memory) as well as Nonvolatile Dual In-line Memory Module (NV-DIMM) requires specific adaptation to the server system somehow in contradiction to the virtualization approach.
- New ways to interact with Flash-based storage not based on block device base methods such as Serial Attached Storage (SAS) but with NVMe (Non-

Volatile Memory Interface) make the difference between memory and permanent storage only visible in access time/latency and achievable bandwidth.

- Specialized hardware solutions with substantially better performance/watt ratio for certain algorithms as standard processors. Even today technology such as General Purpose Graphics processing units (GPGPUs), Manycore solutions, or Floating Point Gate Arrays (FPGAs) are in use within the Data Centre. With them the placement of VMs or tasks is becoming a very complex task and is in particular challenging if an application is supposed to run on many different platforms or is not evolving as fast as the hardware market.

These changes require a visible change in the Cloud Architecture from a model where the hypervisor are hiding the hardware details but emulate to the operating systems or applications toward a model where dynamic runtime optimization can not only happen across applications (similar to what PaaS platform's claim as of today) but also on a much more fine-grained level where tasks within applications can be dynamically moved across the different available resources with much less overhead as currently possible with heavy weighted operating systems. In order to exploit such environments not only a more dynamic runtime is necessary but also applications designed for such a software-defined execution environment.

7.5.2 Latency Optimized Infrastructures

With new Cloud application domain such as optimizing manufacturing processes within an Industry 4.0 environment, the management of traffic flow of a large number of vehicles in an autonomous driving scenario or more general in Internet of Things approaches, a centralized Cloud service provider solution is not sufficient to meet the demands of low-latency communication and response.

In order to address such a scenario, Cloud services must be able to migrate quickly and be driven by optimization rules across different applications between the network edge (with limited storage and compute capacity) and the network core (virtually unlimited storage and compute capacity) and potentially be linked tightly with storage/computing/sensors on the customer side.

Existing platforms partially operate with rather naive placement algorithms for the VMs and rely on manual or simple rules-based migration of VMs across servers typically limited to one single Cloud Data Center. Such an environment is therefore no longer a more classical client–server like architecture with Cloud service consumers and provider but a very distributed system of Cloud service providers distributed across a whole ISP domain delivering certain services at the network edge close to the customer, having one or more Cloud Data Center in the network core potentially with replicated data to support the edge resources, and also interacting with storage and compute resources in a remote network hosted by another ISP. In order to meet quality constraints, not only the performance of all these services must be measured but also optimization algorithms must proactively move services between core and edge or vice versa if the limited resources at the edge are exhausted. This also includes the challenge that a virtual machine or container must be executable in a core data center as well as in potentially very different edge resource.

7.6 CONCLUSION AND SUMMARY

In this chapter we introduced and shortly outlined the typical Cloud layers following the NIST definition to provide a basic understanding of their functionality, using an introduction of the underpinning virtualization technology. Exemplarily alternative or complementing approaches such as Business Process as a Service and Network as a Service had been shortly covered in particular to point out the limitations of the layered model from NIST that by borrowing from the ISO/OSI stack the notion of independent layer does not reflect the typical hybrid solutions that cannot be clearly allocated to one of the layers.

The chapter is concluded by discussing two selected trends that impact the way how future Cloud platforms will have to be built. The virtualization trend started at the time of a rather homogeneous hardware landscape for service platforms is no longer true and will be dramatically changed in the next few years. With this the fundamentals of virtualization of Cloud solutions will change and will have an impact on how the Cloud will be built in the future. The other trend toward distributed Clouds driven by the need to deliver services with low-latency communication, also known as Edge-Computing will also impact how future Clouds will look, particularly with the explosion in the number of supported devices as in the Internet of Things (?).

NOTES

1. Richard Hunter and George Westerman, The real business of it: how CIOs create and communicate business value, *Harvard Business Review*, 2009.
2. NIST definition of Cloud computing. National Institute of Standards and Technology, September 2011 http://csrc.nist.gov/publications/nistpubs/800-145/SP800-145.pdf (last accessed January 7, 2016).
3. Robert Woitsch and Wilfrid Utz, Business process as a service: model-based business and IT Cloud alignment as a Cloud offering. *2015 International Conference on Enterprise Systems*, IEEE, 2015.
4. Paolo Costa et al. NaaS: Network-as-a-Service in the Cloud. Presented as part of the 2nd USENIX Workshop on Hot Topics in Management of Internet, Cloud, and Enterprise Networks and Services. 2012.
5. NIST definition of Cloud computing. National Institute of Standards and Technology, September 2011, http://csrc.nist.gov/publications/nistpubs/800-145/SP800-145.pdf (last accessed January 7, 2016).
6. Steffen Kächele, Christian Spann, Franz J. Hauck, and Jörg Domaschka Beyond IaaS and PaaS: an extended Cloud taxonomy for computation, storage and networking, in *Proceedings of the 2013 IEEE/ACM 6th International Conference on Utility and Cloud Computing (UCC'13)*, IEEE Computer Society, Washington, DC, pp. 75–82. doi: http://dx.doi.org/10.1109/UCC.2013.28.
7. Amazon Web Services announces Amazon Machine Learning. Amazon, http://phx.corporate-ir.net/phoenix.zhtml?c=176060&p=irol-newsArticle&ID=2033665 (April 9, 2015).

8. R.P. Goldberg, Architectural principles for virtual computer systems, 1972.
9. Ana Ferrer et al., OPTIMIS: a holistic approach to cloud service provisioning, *Future Generation Computer Systems*, 2012, 28 (1), 66–77, ISSN 0167-739X, http://dx.doi.org/10.1016/j.future.2011.05.022.
10. Stefan Wesner et al., Optimised Cloud Data Centre Operation Supported by Simulation, eChallenges e-2014 Conference Proceedings, Paul Cunningham and Miriam Cunningham (Eds.), IIMC International Information Management Corporation, 2014, ISBN: 978-1-905824-45-8.

Chapter 8

Emerging Markets: Mobile Money for the Unbanked

As seen in previous chapters, connectivity (with voice progressively disappearing as a standalone service) is the core service offered by telcos and dominated by them (with recent industry consolidation Pay TV is also becoming telco realm). Mobile money in emerging markets is probably the only other service that is observing massive adoption and is starting to contribute significantly to telco revenues in those emerging markets.

Financial inclusion fosters the economy, but still 2 billion adults in the world remain without a bank account.[1,2] This population is concentrated in less developed regions (Africa, part of Asia, and Latin America). A basic use case of sending money to the family for living by a remote worker is quite expensive for people outside the financial inclusion. As commented before, mobile penetration is much more universal and enables these remittances in a cheap and convenient way. End users do not need to have the latest smartphone or mobile broadband, since the technology is based on feature phones and 2G technology (SMS or Unstructured Supplementary Service Data-USSD channel).

Domestic remittance is the most common application. Additional applications include bill payments and merchant payments. International remittances are now also possible even among different operators (cross-MNO agreements). Globally, in 2014, there were already more than 100 million active mobile money accounts, with services deployed in around 90 countries.[3]

8.1 CUSTOMER NEED: REMOTE PAYMENTS

Banks currently underserve in the developing parts of the world, particularly in rural areas, where economics do not make it feasible to have bank branches. Without financial inclusion, people have to travel very long distances (and also wait for a long time in queues) to make financial transactions. Think of use cases such as

Digital Services in the 21st Century: A Strategic and Business Perspective, First Edition.
Antonio Sánchez and Belén Carro.

sending money to the family by remote workers, or paying utility bills. Without banking means, there are not many alternatives and they can be also risky, prone to robbery. Think of people traveling themselves with the money, or having to trust third parties (e.g., public transport drivers) to deliver the money.

Financial inclusion cannot be considered as a goal itself, but rather a means for development. Indeed, a powerful means for poverty reduction. It enables people to save, allowing them to become entrepreneurs, invest in education, absorb financial shocks, and so on.

8.2   LARGE UNBANKED POPULATION IN EMERGING MARKETS

It has already been mentioned that as of 2014, there are approximately 2 billion people without a bank account, that is, unbanked. This figure had decreased significantly over the previous 3 years, a period in which around 700 million entered the financial system by having an account. This number is based on periodic poll starting in 2011. Taking into account population growth and people becoming adults, the number of unbanked people was reduced by half a billion. This is compared to a recent world population figure of 7.2 billion based on a 2012 revision published in 2013 by the United Nations.[4] However, these 2 billion people could improve their lives significantly with financial inclusion, as commented earlier.

The facts shown here come mainly from *Global Findex* (second edition) report from *World Bank* (through its *Research Department*) in collaboration with *Bill and Melinda Gates Foundation* and *Gallup World Poll*, based on a very large survey of almost 150,000 people (15 years old or above, that is, mostly adults) from 143 countries (representing almost the whole world population, 98% to be more precise).

In the 3 years from its first edition, financial inclusion has improved 11 percentage points (from 51 to 62%), and more interestingly 13 percentage points in developing countries. However, financial inclusion among women has not improved as much as it would be desired, since the gender gap (compared to men) is 9 percentage points in developing countries. From a policy perspective, the report forecasts that digitization of salary payments would increase financial inclusion by 300 million people.

Interestingly, mobile money for the unbanked has been one of the main factors contributing to this progress. For example, 13 countries in Sub-Saharan Africa have mobile account penetration equal to or above 10%. Globally the number of adults that only use a mobile money account is 1% (2% the total with mobile money account), including those who have also other accounts, but in Sub-Saharan Africa that figure increases to almost 6% (which in turn is 45% of adults that have mobile money accounts).

Table 8.1 shows for each of the countries in the world (as well as regions, both geographical and economical) the evolution (2011 versus 2014) of percentage of adult (15+) unbanked population, as well as the percentage of mobile account

Table 8.1 Percentage Unbanked, Mobile Account, and Population Per Country[a]

Country name	2011: Unbanked (% age 15+)	2014: Unbanked (% age 15+)	2014: Mobile account (% age 15+)	Population, age 15+ (million)
Turkmenistan	100	98		4
Central African Republic	97			3
Yemen, Rep.	96	94		14
Niger	98	93	4	8
Guinea	96	93	1	6
Burundi	93	93	1	5
Madagascar	94	91	4	12
Afghanistan	91	90	0	15
Iraq	89	89		19
Tajikistan	97	89	0	5
Cameroon	85	88	2	12
Chad	91	88	6	6
Djibouti	88			1
Pakistan	90	87	6	115
Egypt, Arab Rep.	90	86	1	54
Middle East (developing only)	89	86	1	97
Burkina Faso	87	86	3	9
Sudan	93	85		21
Senegal	94	85	6	8
Sierra Leone	85	84	4	3
Benin	90	83	2	6
Congo, Rep.	90	83	2	2
Congo, Dem. Rep.	96	83	9	35
Armenia	83	82	1	2
Moldova	82	82		3
Malawi	83	82	4	8
Togo	90	82	1	4
Kyrgyz Republic	96	82		4
Lesotho	82			1
Haiti	78	81	4	6
Nicaragua	86	81	1	4
Liberia	81			2
Mali	92	80	12	8
Ethiopia		78	0	50
Cambodia	96	78	13	10
Comoros	78			0
Paraguay	78			4
Myanmar		77	0	39
Mauritania	83	77	6	2
Syrian Arab Republic	77			14
West Bank and Gaza	81	76		2

(*continued*)

Table 8.1   *continued*

Country name	2011: Unbanked (% age 15+)	2014: Unbanked (% age 15+)	2014: Mobile account (% age 15+)	Population, age 15+ (million)
Jordan	75	75	0	4
Tunisia		73	1	8
Lao PDR	73			4
Low income	79	72	10	489
Peru	80	71	0	21
Azerbaijan	85	71		7
Angola	61	71		11
Swaziland	71			1
Vietnam	79	69	0	68
Bangladesh	68	69	3	105
Philippines	73	69	4	62
Honduras	79	69	3	5
Zimbabwe	60	68	22	8
Gabon	81	67	7	1
Bhutan		66		1
Nepal	75	66	0	17
Sub-Saharan Africa (developing only)	76	66	12	503
Cote d'Ivoire	100	66	24	11
Zambia	79	64	12	7
Indonesia	80	64	0	172
El Salvador	86	63	5	4
Albania	72	62		2
Somalia		61	37	5
Colombia	70	61	2	34
Mexico	73	61	3	84
Morocco	61			23
Georgia	67	60		4
Tanzania	83	60	32	26
Ghana	71	59	13	15
Uzbekistan	77	59		21
Guatemala	78	59	2	9
Bolivia	72	58	3	7
Rwanda	67	58	18	6
Lower middle income	71	57	2	1695
Panama	75	56	2	3
Nigeria	70	56	2	92
Uganda	80	56	35	18
Uruguay	76	54	1	3
Ecuador	63	54		11
South Asia	68	54	3	1126
Lebanon	63	53	1	3

Table 8.1 *continued*

Country name	2011: Unbanked (% age 15+)	2014: Unbanked (% age 15+)	2014: Mobile account (% age 15+)	Population, age 15+ (million)
Kosovo	56	52		0
Belize		52		0
Argentina	67	50	0	31
Algeria	67	50		27
Latin America and Caribbean (developing only)	61	49	2	414
Europe and Central Asia (developing only)	57	49	0	209
Botswana	70	48	21	1
Bosnia and Herzegovina	44	47		3
Ukraine	59	47		39
India	65	47	2	857
Kazakhstan	58	46		12
Dominican Republic	62	46	2	7
Low and middle income	59	46	3	4042
Turkey	42	43	1	54
Venezuela, RB	56	43	3	21
Middle income	57	42	2	3553
Namibia		41	10	1
Montenegro	50	40		1
Romania	55	39	0	17
World	49	38	2	5097
Bulgaria	47	37		6
Chile	58	37	4	14
Costa Rica	50	35		4
Qatar	34			2
Russian Federation	52	33		121
Brazil	44	32	1	148
East Asia and Pacific (developing only)	45	31	0	1559
Saudi Arabia	54	31		19
Puerto Rico		30		3
South Africa	46	30	14	36
Upper middle income	43	30	1	1858
Macedonia, FYR	26	28		2
Belarus	41	28		8
Hungary	27	28		9
High income: non-OECD	42	27		185
Kuwait	13	27		2
Oman	26			2
Kenya	58	25	58	24

(*continued*)

Table 8.1 *continued*

Country name	2011: Unbanked (% age 15+)	2014: Unbanked (% age 15+)	2014: Mobile account (% age 15+)	Population, age 15+ (million)
Trinidad and Tobago	24			1
Slovak Republic	20	23		5
Poland	30	22		33
Lithuania	26	22		3
Thailand	27	22	1	54
Jamaica	29	22	1	2
China	36	21		1102
Malaysia	34	19	3	21
Bahrain	35	18		1
Czech Republic	19	18		9
Mauritius	20	18	1	1
Sri Lanka	31	17	0	16
Serbia	38	17		6
The United Arab Emirates	40	16	11	8
Croatia	12	14		4
Italy	29	13		51
Portugal	19	13		9
Greece	22	12		10
Israel	10	10		6
Cyprus	15	10		1
Latvia	10	10		2
High income	15	9		1054
Taiwan, China	13	9		0
Mongolia	22	8	5	2
Iran, Islamic Rep.	26	8	4	58
The United States	12	6		250
High income: OECD	10	6		870
Korea, Rep.	7	6		42
Ireland	6	5		4
Euro area	9	5		285
Hong Kong SAR, China	11	4		6
Luxembourg	5	4		0
Malta	5	4		0
Singapore	2	4	6	4
France	3	3		53
Japan	4	3		111
Austria	3	3		7
Slovenia	3	3		2
Spain	7	2		40
Estonia	3	2		1
Switzerland		2		7
Belgium	4	2		9

Table 8.1 *continued*

Country name	2011: Unbanked (% age 15+)	2014: Unbanked (% age 15+)	2014: Mobile account (% age 15+)	Population, age 15+ (million)
Germany	2	1		71
Australia	1	1		18
The United Kingdom	3	1		52
Canada	4	1		29
The Netherlands	1	1		14
New Zealand	1	0		3
Sweden	1	0		8
Denmark	0	0		5
Finland	0	0		5
Norway		0		4

[a]Global Financial Inclusion Database. World Bank, http://databank.worldbank.org/data/reports.aspx? source=1228# (April 15, 2015).
Source: Global Findex (Global Financial Inclusion Database) (last updated April 15, 2015).[25125]

(2014) and the adult population (million 15+). The following observations have been made from Table 8.1:

- In general, and as suggested by global progress, countries have decreased the percentage of unbanked population, but it can be seen that there are still countries with almost all (up to 98%) adult population unbanked.
- There are 82 countries with half or more adult population unbanked.
- In terms of developing geographical regions, developing Middle East has 86% of unbanked population (3 percentage points decrease), and Sub-Saharan Africa 66% (but with very good improvement of 10 percentage points). Other two regions with good decrease are developing Latin America and Caribbean and developing Europe and Central Asia, both with 49% unbanked, decreasing from 61% and 57%, respectively. Finally, developing East Asia and Pacific has improved 14 percentual points to 31%. Globally, in the world (5.1 billion adults) unbanked population is 38% (minus 11 percentage points).
- There are economic areas such as Eurozone (countries that share the Euro currency) or OECD (Organisation for Economic Co-operation and Development) countries labeled as high income, or even in general high-income countries in which the percentage of unbanked is very low (less than 10%).
- In general, countries with unbanked population below 30% do not have mobile money accounts. Most developed countries (northern part of North America with the United States and Canada, and Trinidad and Tobago in the Americas; more developed Asian countries such as Japan, South Korea,

China, Bahrain, Oman, and Kuwait; Australia and New Zealand in Oceania; most of Europe including Western Europe and most of Eastern Europe) belong to this category. There are some exceptions (Singapore and the United Arab Emirates) due to high percentage of immigrant population (explained below). Other exceptions are countries with low unbanked population, but with mobile money services, such as Mongolia (8% unbanked and 5% mobile money penetration) and Iran Islamic Republic (very similar with 8% unbanked and 4% mobile money penetration). Closer to 20 or 25% unbanked are Sri Lanka (17%), Mauritius (18%), Malaysia (19%), Jamaica (22%), and Thailand (22%), all of which report mobile money accounts (Sri Lanka very small with less than 0.1%, the others with 1%, with the exception of Malaysia with 3%). Between 30 and 38% unbanked (which is the world average), there are some countries without mobile money service such as Puerto Rico, Saudi Arabia, Russian Federation, Costa Rica, and Bulgaria (Montenegro is slightly above world average with 40% but did not report mobile money in 2014); but with the same percentage range (30–38%) there are also countries with mobile money, namely, South Africa, Brazil, and Chile. It has to be noted that Kenya only has 25% unbanked, but this is due to highest mobile money penetration (58%), which means that only 17% have a bank account (in other words, unbanked percentage already takes into account mobile money penetration). Similar rationale although not as strong would apply to South Africa (30% unbanked but 14% with mobile money). Above world average, it is quite difficult to find countries without mobile money in 2014, exceptions (besides Montenegro already commented) are Kazakhstan, Ukraine, Bosnia, and Herzegovina, Algeria, Belize, Kosovo, Ecuador (which is a special case given the fact that in 2015 the government is pushing forward its own regulated service), Uzbekistan, Georgia, Albania, Bhutan, Angola, and Azerbaijan (a few more above 75% unbanked such as Moldova, Sudan, Iraq, Yemen, and Turkmenistan); and finally a few others for which no data from the survey are available, including countries such as Paraguay that do have mobile money as will be shown later.

- It has to be noted as well that for countries with relevant mobile money penetration (e.g., 10% of above), the decrease of unbanked population in the period 2011–2014 has been also relevant, with mobile money contributing significantly to that improvement. To show a few examples: Mali has improved from 92 to 80%, and mobile money is precisely 12%; Cambodia from 96 to 78%, with mobile money 12%; Cote d'Ivoire from 100 to 66%, with mobile money 24%; Zambia from 79 to 64%, with mobile money 12%, Tanzania, Ghana, Rwanda, and Uganda decreasing 23, 12, 9, and 24%, respectively, and mobile money penetration of 32, 13, 18, and 35%, respectively.

- It is worth mentioning that the definition of mobile money penetration only considers usage in the past 12 months, in other words, it requires active usage, but the period in order to consider activity is relatively long. On the other hand, the unbanked penetration just considers having an account, without explicit mention that it is active.

A total of 74 countries report mobile money accounts, and the number is growing with new countries launching services. As expected, it can be seen that the country with highest percentage of penetration of mobile money accounts is Kenya with 58%, followed by Somalia (37%), Uganda (35%), Tanzania (32%), Cote d'Ivoire (24%), Zimbabwe (22%), and Botswana (21%). It can be seen that the top four countries with a penetration above 30% are located in West Africa, all located together around Kenya (the pioneer) with which they share a border. The next three, which are above 20%, are also in the African continent, the first in the East and the last two in the South but close to the first four (especially Tanzania). Additional eight countries have a penetration equal to or above 10% (Rwanda 18%, which could also be considered as part of the West Africa cluster, and between 10 and 15% for South Africa, Cambodia, Ghana, Zambia, Mali, the United Arab Emirates, and Namibia), also including countries in Asia. With penetration between 5 and 10%, there are nine more countries, including El Salvador (5%)—the first one in Latin America. In general, all these countries share a low Gross Domestic Product (GDP) per capita (e.g., Kenya, Uganda, Tanzania, Zimbabwe, Rwanda, Cambodia, Ghana, Zambia, Mali—about $2000–$4000, Cote d'Ivoire $7000, Namibia $10,000. In general, most countries with mobile money penetration equal to or above 10% have just a few thousand dollars GDP per capita and below $10,000 threshold; same rule of thumb would apply to next group of countries in the list including Congo Democratic Republic, Mauritania, Senegal, Pakistan, Chad, and El Salvador, the latter with $8000). Exceptions are wealthy countries such as the United Arab Emirates or Singapore (third and sixth countries in the world by GDP per capita), with large immigrant population, that are the target users of mobile money for unbanked, sending international remittances to their families who are unbanked in other less developed countries. Other exceptions could be Botswana, whose GDP per capita (with purchasing power parity) is approximately $17,500[5] (current international dollar), Gabon with $19,000, South Africa with $13,000, and Mongolia with $12,000. All this information is compiled in Table 8.2.

Statistics show additional granularity in terms of mobile money accounts demographics, as can be seen in Table 8.3. Taking the top 10 countries by mobile money penetration (which are all African plus Cambodia which is the 10th), and adding the first country from the Americas (El Salvador 24th), we can have a look at details such as age, education, income, gender, and rural condition.

One of the very first observations is that education is a key influencing factor, since in all countries the penetration among people with secondary education or more is higher than the average, in some case almost twice as much (e.g., in Somalia 65 versus 37% total and 32% among those with primary education or less).

Table 8.2   GDP Per Capita of Countries with Largest Mobile Account Penetration

Country name	2014: Mobile account (% age 15+)	GDP per capita (PPP international $)
Kenya	58	3
Somalia	37	
Uganda	35	2
Tanzania	32	3
Cote d'Ivoire	24	7
Zimbabwe	22	2
Botswana	21	18
Rwanda	18	2
South Africa	14	13
Cambodia	13	3
Ghana	13	4
Zambia	12	4
Mali	12	2
The United Arab Emirates	11	63
Namibia	10	10
Congo, Dem. Rep.	9	6
Gabon	7	19
Mauritania	6	4
Senegal	6	2
Singapore	6	83
Pakistan	6	5
Chad	6	2
Mongolia	5	12
El Salvador	5	8

In terms of age, in several countries older (25+) people use it more, but it cannot be generalized. In terms of income, penetration is higher in richer segment, which appears natural as first subsegment among unbanked to get adoption. Finally, there is more penetration among male, which hints about less access to financial inclusion among women.

In terms of rural penetration, with the exception of Kenya (56%), all other countries shown have a penetration below 40%. Arguably, it could be expected that rural penetration could be high, given the lack of bank branches, and the long distances required to make financial transactions, where mobile money's unmet customer need is more relevant.

Back to education, it has to be noted that mobile money benefits from literacy, since illiterate people have difficulty using a mobile phone through its screen (inability to read). That is why some services are based on voice interfaces. Although literacy rates are increasing and are already high in many countries, there are still almost 50 countries (for which data are available) in the world

Table 8.3 Mobile Money Account Demographics for Countries with Largest Penetration

Mobile account 2014 (%, age 15+)	Total	Rural	Female	Male	Income, poorest 40%	Income, richest 60%	Young adults (% ages 15–24)	Older adults (% ages 25+)	Primary education or less	Secondary education or more
Kenya	58	56	55	62	53	62	52	62	51	67
Somalia	37	31	32	42	26	45	41	36	32	65
Tanzania	32	28	27	38	17	42	26	36	26	56
Uganda	35	36	29	41	21	45	27	41	28	46
Zimbabwe	22	18	19	24	11	29	20	23	13	27
Cote d'Ivoire	24	26	20	28	20	27	20	26	21	34
Botswana	21	21	19	22	10	28	23	20	5	29
South Africa	14	13	14	15	8	19	10	16	6	18
Rwanda	18	16	16	20	8	26	12	21	16	30
Cambodia	13	13	13	14	11	15	18	11	12	20
El Salvador	5	6	4	5	4	5	4	5	4	6

with adult literacy rates below 75% (and around 75 with rates below 90%),[6] as can be seen in Table 8.4. It has to be noted that successful mobile money markets such as Kenya, Uganda, and Cambodia have literacy rates of 77–78%, Zimbabwe, Botswana, and El Salvador 87–88%, and South Africa 94%

Table 8.4   Countries with Lowest (Below 75%) Adult Literacy Rate in 2015

Country	Adult (15+) literacy rate in 2015 (%)
Niger	19
Guinea	30
Burkina Faso	36
Central African Republic	37
Afghanistan	38
Benin	38
Mali	39
Chad	40
Côte d'Ivoire	43
Liberia	48
Sierra Leone	48
Ethiopia	49
Mauritania	52
Gambia	56
Senegal	58
Pakistan	58
Mozambique	59
Nigeria	60
Guinea-Bissau	60
Haiti	61
Bangladesh	62
Zambia	63
Democratic Republic of the Congo	64
Nepal	64
Papua New Guinea	64
Madagascar	65
Bhutan	65
Malawi	66
Togo	67
Timor-Leste	68
Morocco	68
Yemen	70
Rwanda	71
The United Republic of Tanzania	71
Angola	71
India	71
Egypt	74
Eritrea	74

(no data for Somalia is available). In Tanzania and Rwanda, it is 71%. Cote d'Ivoire is the only outlier among countries with more mobile penetration with only 43% literacy rate.

8.3 VERY HIGH PENETRATION OF MOBILE BASED ON FEATURE PHONES

In the world there are already more mobile subscriptions than people, with 7.5 billion mobile subscription threshold surpassed in July 2015.[7] It has to be noted however that not all lines belong to people (some are machine to machine lines), and there are people with more than one line: different lines for corporate versus personal use, different lines from different operators to have access to different tariffs. Unique mobile subscribers are estimated to be approximately half of that figure, that is, 3.7 billion. On the other hand, it is also worth mentioning that there are population segments without mobile, be it elderly people (but this segment tends to disappear) or children, which might not have a mobile subscription until they become adults or teenagers, being difficult to establish a threshold age. Taking into account these considerations, Table 8.5

Table 8.5   Mobile Penetration Compared to Financial Account Penetration

Country name	Mobile cellular subscriptions (per 100 people)	2014: Bank account (% age 15+)	2014: Mobile account (% age 15+)	Differential: mobile versus bank accounts
Afghanistan	71	10	0	61
Albania	116	38		78
Algeria	101	50		50
Angola	62	29		33
Argentina	163	50	0	112
Armenia	112	18	1	95
Australia	107	99		8
Austria	156	97		59
Azerbaijan	108	29		78
Bahrain	166	82		84
Bangladesh	74	31	3	43
Belarus	119	72		47
Belgium	111	98		13
Belize	53	48		4
Benin	93	17	2	77
Bhutan	72	34		39
Bolivia	98	42	3	56
Bosnia and Herzegovina	91	53		38

(*continued*)

Table 8.5 *continued*

Country name	Mobile cellular subscriptions (per 100 people)	2014: Bank account (% age 15+)	2014: Mobile account (% age 15+)	Differential: mobile versus bank accounts
Botswana	161	52	21	109
Brazil	135	68	1	67
Bulgaria	145	63		82
Burkina Faso	66	14	3	52
Burundi	25	7	1	18
Cambodia	134	22	13	112
Cameroon	70	12	2	58
Canada	81	99		−18
Central African Republic	29			NA
Chad	36	12	6	23
Chile	134	63	4	71
China	89	79		10
Colombia	104	39	2	65
Comoros	47			NA
Congo, Dem. Rep.	42	17	9	24
Congo, Rep.	105	17	2	88
Costa Rica	146	65		81
Cote d'Ivoire	95	34	24	61
Croatia	115	86		28
Cyprus	96	90		6
Czech Republic	128	82		46
Denmark	127	100		27
Djibouti	28			NA
Dominican Republic	88	54	2	34
Ecuador	111	46		65
Egypt, Arab Rep.	122	14	1	107
El Salvador	136	37	5	99
Estonia	160	98		62
Ethiopia	27	22	0	5
Finland	172	100		72
France	98	97		2
Gabon	215	33	7	182
Georgia	115	40		75
Germany	121	99		22
Ghana	108	41	13	68
Greece	117	88		29
Guatemala	140	41	2	99
Guinea	63	7	1	56
Haiti	69	19	4	51
Honduras	96	31	3	64

Table 8.5 *continued*

Country name	Mobile cellular subscriptions (per 100 people)	2014: Bank account (% age 15+)	2014: Mobile account (% age 15+)	Differential: mobile versus bank accounts
Hong Kong SAR, China	237	96		141
Hungary	116	72		44
India	71	53	2	18
Indonesia	125	36	0	89
Iran, Islamic Rep.	84	92	4	−8
Iraq	96	11		85
Ireland	103	95		8
Israel	123	90		33
Italy	159	87		71
Jamaica	102	78	1	24
Japan	118	97		21
Jordan	142	25	0	117
Kazakhstan	185	54		131
Kenya	72	75	58	−3
Korea, Rep.	111	94		17
Kosovo		48		NA
Kuwait	190	73		117
Kyrgyz Republic	121	18		103
Lao PDR	68			NA
Latvia	228	90		138
Lebanon	81	47	1	34
Lesotho	86			NA
Liberia	59			NA
Lithuania	151	78		73
Luxembourg	149	96		52
Macedonia, FYR	106	72		34
Madagascar	37	9	4	28
Malawi	32	18	4	14
Malaysia	145	81	3	64
Mali	129	20	12	109
Malta	130	96		33
Mauritania	103	23	6	80
Mauritius	123	82	1	41
Mexico	86	39	3	47
Moldova	106	18		88
Mongolia	124	92	5	32
Montenegro	160	60		100
Morocco	129			NA
Myanmar	13	23	0	−10
Namibia	118	59	10	60
Nepal	77	34	0	43

(continued)

Table 8.5   *continued*

Country name	Mobile cellular subscriptions (per 100 people)	2014: Bank account (% age 15+)	2014: Mobile account (% age 15+)	Differential: mobile versus bank accounts
The Netherlands	114	99		14
New Zealand	106	100		6
Nicaragua	112	19	1	93
Niger	39	7	4	33
Nigeria	73	44	2	29
Norway	116	100		16
Oman	155			NA
Pakistan	70	13	6	57
Panama	163	44	2	119
Paraguay	104			NA
Peru	98	29	0	69
The Philippines	105	31	4	73
Poland	149	78		71
Portugal	113	87		26
Puerto Rico	84	70		14
Qatar	153			NA
Romania	106	61	0	45
Russian Federation	153	67		85
Rwanda	57	42	18	15
Saudi Arabia	184	69		115
Senegal	93	15	6	78
Serbia	119	83		36
Sierra Leone	66	16	4	50
Singapore	156	96	6	60
Slovak Republic	114	77		37
Slovenia	110	97		13
Somalia	49	39	37	11
South Africa	146	70	14	75
Spain	107	98		9
Sri Lanka	95	83	0	13
Sudan	73	15		58
Swaziland	71			NA
Sweden	124	100		25
Switzerland	137	98		39
Syrian Arab Republic	56			NA
Tajikistan	92	11	0	80
Tanzania	56	40	32	16
Thailand	140	78	1	62
Togo	63	18	1	44
Trinidad and Tobago	145			NA

Table 8.5 *continued*

Country name	Mobile cellular subscriptions (per 100 people)	2014: Bank account (% age 15+)	2014: Mobile account (% age 15+)	Differential: mobile versus bank accounts
Tunisia	116	27	1	88
Turkey	93	57	1	36
Turkmenistan	117	2		115
Uganda	44	44	35	0
Ukraine	138	53		85
The United Arab Emirates	172	84	11	88
The United Kingdom	125	99		26
The United States	96	94		2
Uruguay	155	46	1	109
Uzbekistan	74	41		34
Venezuela, RB	102	57	3	45
Vietnam	131	31	0	100
West Bank and Gaza	74	24		49
Yemen, Rep.	69	6		63
Zambia	72	36	12	36
Zimbabwe	96	32	22	64

compares the penetration of mobile versus financial account in the different countries for which data are available.[8] Just illustratively and as some sort of proxy (given that one metric refers to the whole population, while the other refers only to adults), the difference between both is shown. In practically all cases, there are more mobile subscriptions than bank accounts. It has to be noted that in Kenya, financial accounts have grown considerably, thanks to mobile money, with majority of people having both, and the remaining people either having mobile without financial account or bank account but not mobile.

It has to be noted that most of phones, particularly in emerging markets, are feature phones, not smartphones. Forecast estimates the installed base of the latter will be only 3 billion in 2016.[9] Therefore, mobile money services have been designed to work in all mobile phones, namely, in feature phones (although applications for smartphones are emerging).

So far, statistics about the demand side (e.g., population) have been shown. In the following paragraphs, figures from the offer side (mainly telecommunications operators) will be shown.

Provider (second quarter 2015 data if available)	Revenues	Users	Average revenue per user (monthly)	Percentage of mobile revenues	Transactions	Markets[10]
Millicom[11]	$31 million (+23% year-over-year)	10.2 million customers (+27% year-over-year)	1 $	3% of mobile service revenues (9% in Africa)		Latin America (El Salvador, Guatemala, Honduras, Bolivia, Paraguay) and Africa (Chad, Democratic Republic of Congo, Ghana, Rwanda, Senegal, Tanzania)
Vodafone[12,13,14]	Kshs 32.6 billion, +23% year-over-year in Kenya, +16.1% Vodacom	501,000 active customers in India. 7.8 million customers (active 90-day) with 18.1% growth in Vodacom subsidiary (of which 5.6 million active 30-day in Tanzania, Democratic Republic of Congo, Mozambique, Lesotho); 13.9 million active (30-day, 60% of total customer base) and 20.6 million registered in Kenya (2015 first quarter) +14% year-over-year	Kshs 209 (30-day active)	21% of mobile service revenues in Kenya (+2 percentage points year-over-year). 9.4% of service revenue (fourth quarter 2014) in Vodacom subsidiary except South Africa.[15] 21.3% of service revenue in Tanzania (second quarter 2014)		Asia (India), Africa (Egypt; Safaricom Kenya; Vodacom: South Africa, Tanzania, Democratic Republic of Congo, Mozambique, Lesotho), Oceania (Fiji), and Europe (Romania and Albania)
MTN[16]	2014: 6.5 million active subscribers 2015 first quarter: 27.4 million registered subscribers, up 23% quarter-over-quarter across 14 operations			2015 first quarter: 5% of total revenue in Ghana. Uganda key market. Ivory Coast and Rwanda also good traction. 2014 first half: Uganda Mobile Money up to 14.7% of Ugandan revenue		Africa (Benin, Cameroon, Congo, Côte d'Ivoire, Ghana, Guinea, Guinea-Bissau, Rwanda, South Africa, Sudan, Swaziland, Uganda, Zambia: Nigeria bank partnership and brand)

Company	Revenue/financials	Customers	% of revenues	Transactions	Geography
Airtel[17]		First quarter 2015 in Africa: 6.2 million customer base using the service (+76.4% year-over-year)		First quarter 2015 in Africa: 157 million transactions (+72.2% year-over-year). $2.9 billion value of transactions (+31% year-over-year)	Africa (Burkina Faso, Chad, Congo, Democratic Republic of Congo, Gabon, Ghana, Kenya, Madagascar, Malawi, Niger, Nigeria, Rwanda, Sierra Leone, Seychelles, Tanzania, Uganda, Zambia) and Asia (India and Bangladesh)
Telenor[18,19]	Pakistan 2014: NOK 560 millions (15Q2: *continued growth in financial services (EBITDA margin decreased[...] increased commissions and taxes for financial services)*	Close to 13 million customers	Pakistan 2014: 9% of total revenues	Pakistan 2014: 113 million money transfers/total transfer value US$3 billion (+42% year-over-year)	Asia (Pakistan, Bangladesh)
Orange[20]	+76% revenues year-over-year (first quarter 2015: +84% revenues year-over-year)	14.2 million customers (+37% year-over-year). (first quarter 2015: 13.3 million customers (+48% year-over-year))			Africa (Botswana, Cameroon, Côte d'Ivoire, Guinea, Kenya, Madagascar, Mali, Mauritius, Niger, Senegal, Uganda), Asia (Jordan)
Econet Wireless[21]	Year ended February 28, 2015: US$60 million	February 2015: 4.2 million (+20% year-over-year) (77% mobile money subscriber share, 98% mobile money deposits share)	8% revenue contribution	February 2015: US $5.5 billion (+77% year-over-year)	Africa (Zimbabwe)
America Movil	Not disclosed				Americas (Mexico, Brazil, Colombia, Guatemala)
Telefonica	Not disclosed				Americas (Brazil, Peru)
Axiata	Not disclosed				Asia (Bangladesh, Indonesia, Malaysia, Sri Lanka)

(*continued*)

(Continued)

Provider (second quarter 2015 data if available)	Revenues	Users	Average revenue per user (monthly)	Percentage of mobile revenues	Transactions	Markets[10]
Etisalat	Not disclosed					Asia (Afghanistan, the United Arab Emirates), Africa (Benin, Côte d'Ivoire, Egypt, Gabon, Niger, Tanzania, Togo, Morocco)
Ooredoo	Not disclosed					Asia (Qatar), Africa (Tunisia)
Zain	Not disclosed					Asia (Jordan)
Telesom[22]		450,000 registered users (31% women with 59% active rate, versus 79% male active) by June 2014			52 and 38 transactions per month by men, women, respectively (June 2014)	Africa (Somaliland)
Digicel	Not disclosed					Americas (Haiti), Oceania (Fiji, Papua New Guinea, Samoa, Tonga, Vanuatu)

In the beginning of 2015, there were around 150 mobile money services from telecommunications operators (out of a total of slightly above 250). As expected, the region with more services is Sub-Saharan Africa, where more than half of mobile network operators have deployed a service, and 23% of mobile lines, were linked with a mobile money account. In 2014, there were a total of almost 300 million registered accounts, which means an activity rate slightly above 33% (but it has to be taken into account that users might be active but only making transactions from time to time). The growth potential is high, given the fact that 300 million represents only 8% of the total of mobile lines in the markets where the service is available. It is noteworthy mentioning the statistics that there are 16 markets in which there are more mobile money accounts than bank accounts. Among the 103 million active mobile money accounts, there are 21 services with more than 1 million active accounts. In addition, there were in 2014 (month of June) 33 million unregistered customers that made over-the-counter transactions. These customers could also be considered as active, the difference is that they perform the transactions through an agent in an outlet. This is a model that is also valid, and is working well in some markets (main example is probably Pakistan, with Telenor and other providers). One of its advantages is that need to educate customers to use this service is reduced.

One of the key success factors of mobile money is distribution capillarity, with mobile money agents, which play a crucial role in attracting customers and educating them. As expected, in most (75%) mobile money markets there are already more agent outlets than bank branches. In a few markets (25) the proportion is higher than 10 to 1.

Besides Africa, it is interesting to see how the service is having success elsewhere. As a matter of fact, in Latin America and the Caribbean, the number of active accounts increased 50% over the year.

Although high percentage of unbanked population and high mobile penetration are two typical ingredients of mobile money, a very critical element to enable it is that of regulation. Fortunately, it has been evolving positively over the past years, but there are still markets that without a level playing field for nonbanks to operate, services cannot take off. It could be said that approximately in half of the markets where there is a live mobile money service, suitable regulation is in place, allowing both banks and nonbanks to provide mobile money services without regulatory hurdles. As an example, Colombia approved in 2014 a law for financial inclusion (*Companies Specialized in Deposits and Electronic Payments*); and in mid-2015 the country approved the normative that details the legislation, including topics such as requirements for capital, cash handling, know-your-customer simplified procedures for monthly balances not exceeding three times the minimum salary, and permission to operate through agents.[23]

8.4 SERVICES: REMITTANCES AND PAYMENTS

Analysts estimated $2 billion revenues from mobile money services in 2015.[24] This is based on the fees that are paid by the customers for the transactions performed.

It could be said that the basic building block of a mobile money service, or in other words the main service, is that of the national peer-to-peer money transfer. As already commented, the customer needs to send money to another person (e.g., a relative) who lives far away. For a small fee, the money can be sent in an easy, safe, and convenient way to the destination in electronic form. A variant of this service is to send the money abroad. This is typically a service that is offered by the provider later on. From the customer perspective, the user experience is very similar, but from the provider perspective, the complexity can be much higher, since it might involve different regulations in home and destination country, foreign exchange, and so on. However, the cost improvements for international remittances are quite good compared to existing alternatives, which can be decreased to less than half (e.g., 4 dollars out of a remittance of 100). In the beginning, these remittances (national or international) were restricted to customers that belong to the same provider. However, there is a strong push for interoperability among providers, both national and international. The first market with national interoperability was Indonesia in 2013, followed by Pakistan, Sri Lanka, and Tanzania in 2014. For these peer-to-peer transactions, such as in banking systems, it is usually the one who sends the money who pays the commission.

Another set of services are those that offer transactions between people and companies, in either direction. People can pay bill through mobile money, being typical the case of bills from utilities (water, electricity, etc.). Without the service, people had to travel to the office of the utility, sometimes even having to queue for very long time in order to make the payment on a periodical basis. Another use case is that of merchant payments. For these ones, typically the customer is in the shop to get the product, but the service offers an alternative way of paying without the need for carrying cash. In 2014, there were around a quarter of a million of merchants that accepted mobile money payments, although only a quarter of them were active. In these cases, like with other payment methods such as credit cards, it is the company (bill issuer, merchant) that pays the transaction fee.

A special case of utility bills is that of telecommunications, and top ups to increase the balance available in the prepaid mobile phone account (being prepaid lines predominant in these markets). Since typically mobile money services are offered by mobile network operators, it is natural to allow customers to use mobile money to buy airtime credit. It is worth mentioning that for operators it is a much cheaper way of doing so, given the higher commissions they have to pay to distributors of mobile airtime balance (e.g., through scratch cards).

In the opposite direction, people can receive salaries from companies or other welfare incomes (social protection, retirement pensions) from governments through mobile money. In the latter case, and for large companies, these can be considered bulk payments or disbursements. In December 2014, transactions involving companies represented 24% by transacted value.

In the previous paragraphs, the basic portfolio of services has been covered. Two companion transactions that complement the systems are cash-in and cash-out operations. The former constitutes the typical way of introducing money in the account. This transaction is typically free, since providers want to encourage customers to have money in their balances, as a prior requirement to get revenues from future transactions. On the other hand, cash-out removes money from the system, and it has some sort of penalty, making users to pay a fee for doing so. In the initial stages of a mobile money service, the ecosystem is not much developed, and there are not many merchants or utilities that accept mobile money, therefore when a person receives money and needs to spend it, it has to cash it out before doing so. But in more mature ecosystems, ideally all financial transactions could be done within the mobile money domain.

In terms of usage, in 2014 top-up (more than 60%) was the product most used in terms of number of transactions, whereas peer-to-peer transfer the one in terms of value of transactions (almost 75%, with 25% of volume).

Finally, for the most mature services, it is natural to resemble the offerings of banks, with additional features such as savings (that accrue interest) and loans.

Other related service is that of mobile insurance (e.g., life insurance), which is complementary to mobile money but can be operated outside of it. In 2014, there were 100 mobile insurance services in 30 markets, totaling 17 million policies for the unbanked. Like mobile money, it decreases significantly the costs to operate, enabling a wider reach.

ACRONYMS

EBITDA	Earnings Before Interests Taxes Depreciation and Amortization
GDP	Gross Domestic Product
Kshs	Kenyan Shillings
MNO	Mobile Network Operator
NOK	Norwegian Krone
SMS	Short Messaging System
USSD	Unstructured Supplementary Service Data

NOTES

1. Demirguc-Kunt, Asli, Leora Klapper, Dorothe Singer, and Peter Van Oudheusden. The Global Findex Database 2014: measuring financial inclusion around the world. Policy Research Working Paper 7255, World Bank, Washington, DC, http://data.worldbank.org/data-catalog/financial_inclusion (April 15, 2015).
2. Global Findex 2014 unveils world's most comprehensive set of data on financial inclusion. World Bank, //www.worldbank.org/en/news/feature/2015/04/20/global-findex-2014-unveils-worlds-most-comprehensive-set-of-data-on-financial-inclusion (April 21, 2015).
3. New 2014 State of the Industry Report on Mobile Financial Services for the Unbanked. Claire Scharwatt, GSMA, //www.gsma.com/mobilefordevelopment/new-2014-state-of-the-industry-report-on-mobile-financial-services-for-the-unbanked (March 3, 2015).
4. World population projected to reach 9.6 billion by 2050 with most growth in developing regions, especially Africa – says UN. *World Population Prospects: The 2012 Revision*, the United Nations, http://esa.un.org/wpp/Documentation/pdf/WPP2012_Press_Release.pdf (June 13, 2013).

5. GDP per capita, PPP (current international $). World Bank, 2014. http://data.worldbank.org/indicator/NY.GDP.PCAP.PP.CD

6. Education: literacy rate. United Nations Educational, Scientific and Cultural Organization (UNESCO) Institute for Statistics, http://data.uis.unesco.org/Index.aspx?queryid=166 (data extracted July 24, 2015).

7. Mobile connections, including M2M/unique mobile subscribers. GSMA Intelligence, https://gsmaintelligence.com/ (July 24, 2015).

8. Mobile cellular subscriptions (per 100 people). The World Bank, http://data.worldbank.org/indicator/IT.CEL.SETS.P2?order=wbapi_data_value_2013+wbapi_data_value+wbapi_data_value-last&sort=asc (last updated July 1, 2015).

9. Three billion global smartphone installed base will be achieved in 2016. Strategy Analytics, //www.strategyanalytics.com/strategy-analytics/blogs/smart-phones/2015/07/01/three-billion-global-smartphone-installed-base-will-be-achieved-in-2016#.VbJPbNLtlBc (July 1, 2015).

10. MMU Deployment Tracker: featuring 265 live deployments and 102 planned deployments. GSMA, //www.gsma.com/mobilefordevelopment/programmes/mobile-money-for-the-unbanked/insights/tracker (last accessed August 1, 2015).

11. Investors. Millicom. //www.millicom.com/media/2940894/good-performance-focus-on-profitable-growth.pdf //www.millicom.com/media/2940892/q2-2015-results-millicom-final-presentation.pdf //www.millicom.com/media/2940890/financial-and-operational-data-q2-15.xlsx //www.millicom.com/investors/ (July 21, 2015).

12. Investor Relations, quarter ended 30th June 2015. Vodafone, Vodacom, //www.vodafone.com/content/dam/vodafone/investors/financial_results_feeds/ims_quarter_30june2015/q1-15-16-trading-update.pdf //www.vodacom.co.za/cs/groups/public/documents/document/trading-statement-230715.pdf (July 24, 2015).

13. Investor Relations: full year results announcement, 2014/2015 FY. Safaricom, //www.safaricom.co.ke/financial-updates/half-year-and-full-year (May 6, 2015).

14. Quarterly update for the period ended 30 June 2014. Vodacom Group Limited, //www.vodacom.co.za/cs/groups/public/documents/vodacom.co.za_portal_webassets/quarterly_update_30_jun_2014.pdf (July 24, 2014).

15. Quarterly update for the period ended 31 December 2014. Vodacom Group Limited, //www.vodacom.co.za/cs/groups/public/documents/vodacom.co.za_portal_webassets/announcement-040215.pdf (February 4, 2015).

16. Interim results 2014, 2015/Quarterly Trading Update 2015 Q1. MTN, //www.mtn.com/Investors/FinancialReporting/Pages/InterimResults.aspx, //www.mtn.com/Investors/FinancialReporting/Documents/INTERIMREPORTS/2014/Booklet/MTN_Interim_Results_booklet_Aug__2014.pdf / //www.mtn.com/Investors/FinancialReporting/Pages/QuarterlyResults.aspx (August 5, 2015).

17. Quarterly results. Airtel, //www.airtel.in/about-bharti/investor-relations/results/quarterly-results/ (August 4, 2015).

18. Q4/ 2014 Interim report. Telenor, //www.telenor.com/wp-content/uploads/2014/09/Telenor-Q4-report-2014-110215.pdf, //www.telenor.com/investors/reports/2015/telenors-results-for-the-4th-quarter-2014/ (February 11, 2015).

19. Telenor Group's results for the 2nd quarter 2015. Telenor, //www.telenor.com/investors/reports/2015/telenors-results-for-the-2nd-quarter-2015/ (July 22, 2015).

20. Investor Relations. Orange, //www.orange.com/en/investors (July 28, 2015) //www.orange.com/en/content/download/30282/842655/version/4/file/Q1+2015+presentation+EN+sans+script+-+vDEF2.pdf (April 28, 2015).

21. Audited results and presentation for FY ended 28 February 2015. Econet Wireless, //www.ewzinvestor.co.zw/profiles/investor/ResLibraryView.asp?ResLibraryID=77242&BzID=1685&t=1402&g=604&Nav=0&LangID=1&s=0 (May 21, 2015).

22. Reaching half of the market: women and mobile money—the example of Telesom in Somaliland. GSMA connected women. //www.gsma.com/connectedwomen/reaching-half-of-the-market-women-and-mobile-money-the-example-of-telesom-in-somaliland/ (April 27, 2015).

23. Gobierno expide decreto reglamentario de la Ley de Inclusión Financiera. Treasury Ministry, Government of Colombia, //www.minhacienda.gov.co/HomeMinhacienda/saladeprensa/07152015-comunicado-72 (July 15, 2015).

24. Africa leads mobile money boom as service providers benefit from $2 bn revenue opportunity. Juniper Research, //www.juniperresearch.com/press/press-releases/africa-leads-mobile-money-boom-2bn-opportunity (June 23, 2015).

Chapter 9

Value-Added Consumer Services

Jesus Llamazares Alberola

9.1 INTRODUCTION

Talking about VAS is always a challenge, as it is a business under dynamic change. This is something you start writing with an initial picture in your head and you observe changes and new on/off stuff even before you finish. It is in constant motion and evolution. That's part of the game, players need to be quick testing, adopting, succeeding and failing, and not fear to make investments on initiatives not succeeding. Of course, the company DNA is very relevant on how to face these new challenges, but we should always be prepared to disrupt service, workflows, legacies, and so on, everything in order to succeed and sometimes survive.

VAS are top of mind when talking about customer needs and customer insights. Therefore, I will write my vision from two decades of experience in the consulting and Technology, Media and Telecommunications (TMT) industry rather than deep diving into technical stuff you can find in other writings. I want to take the readers through the TMT "tides," reflect together on why some telcos survived and others not or how other players out of telco industry in the last decade appeared to succeed and even lead some part of telcos value chain. Disruption is no longer an extremist aphorism, it is being hard wired in successful companies' DNA and we all should think and adopt it as part of our life.[1]

9.2 DISRUPTION IS THE NEW "KARMA"

In the early 1980s, telco's portfolio was extremely simple: fixed line and fixed voice was the core service. Sales representatives enjoyed barely one KPI: number of connected lines per month. So business was quite easy to manage, everything was about fixed (copper) network planning and coverage where in most countries an incumbent was the one solely responsible for this task, with no competition.

Digital Services in the 21st Century: A Strategic and Business Perspective, First Edition.
Antonio Sánchez and Belén Carro.
© 2017 by The Institute of Electrical and Electronics Engineers, Inc. Published 2017 by John Wiley & Sons, Inc.

Customer operations were just focused on installation, activation, and billing claims. When we look back, it seems that this "simple" business was running ages ago, but we are just talking about three decades ago.

Then some advanced (at that time) fixed VAS appeared: voicemail, re-dial, even call history and agenda with some DECT advanced phones. Telcos charged some extra for some of these services and got a new revenue stream, at least for some time.

In the early1990s, two new phenomena appeared: fixed line Internet and initial mobile telephony stages. Let us talk first about Internet and connection to corporate networks through fixed lines. In the very early stage, we connected our PCs to telephone lines (without splitters) to connect to the Internet and corporate networks through dial-up calls. It was not broadband speed but was just enough to check your email or, having lots of patience, connect to some corporate system or surf some simple webs. You really used it for key relevant things as you spent lots of time in doing so. And complexity was high enough as to install clients in PC that managed country numbers, type of connections, and so on. Telcos focused billing in the same way as regular calls, based on minutes to premium numbers, where part of the service was a revenue share with the Internet provider.

Mobile industry can be considered as the first disruption in telecommunications. We can read tons of books about evolution of technology and services, but what I would like to share with you is that at the very early moments, mobile business started as a very small department within fixed line incumbent telco giants. With 4–5 kg devices costing 3000+ euros in the late 1980s and lack of network coverage in many cities and areas, business seemed to be neither massive nor successful as a new big revenue stream for telcos. Many traditional home devices OEMs were not prepared (operationally, strategically, or both) to jump into this new business and, thus, only network manufacturers such as Ericsson and Nokia were offering these new products. In the early 1990s, other OEMs, no network focused, joined the business (mainly from Southeast Asia) and dropped prices down. These new prices, jointly with massive subsidies from telcos and distributors, drove a massive adoption of the service. What was in the beginning a small department of 5–30 employees soon became 20–30% of the total revenues. Though evolution of mobile industry was faster paced than fixed line industry, we started to suffer a staggering in revenue projections in early new millennium when SMS became the most important VAS accruing up to 10–15% of the revenue, although with no vision to get better figures.

Meanwhile, in the United States, two major forces were rising: Microsoft and Apple. Initially fighting for the best OS, they had a fundamental difference between them. Microsoft was just focused on OS and work apps, and Apple soon understood the "magic" of building hardware and software at the same time and in connection, getting the best jointly, and mainly focused on entertainment, creativity, and design. This race achieved a big milestone in the TMT industry when Apple launched its first iPhone, Apple Store, and iTunes in 2006. This was one of the biggest disruption moves ever. The full launch of a complete ecosystem based on best HW-SW performance, access to features through Apps rather than accessing webs, and access to lots of media contents was the start of a game change in tele-communications industry. Internet players jumped into the App world so rapidly that as of today 90%[2] of time spent in smartphones is on Apps. This milestone triggered the creation of a multilayered value chain.

9.3 ADJACENT INDUSTRIES JOINING MULTILAYERED VALUE CHAIN

Before Apple announced iPhone in 2006, value chains were clearly defined: telco, financial services, media, and so on. And players from different industries enjoyed small or no cross-industry competition. Before 2006, there were just few Internet examples such as Google in advertising and Paypal in financial services that started shyly to challenge established industries. At this point incumbents did not feel the challenge nor react predicting the future.

Ten years after, we can see a quite different picture. Let us fly through some examples:

- *Financial services*
  - Initiatives such as Square, Paypal, and Bitcoin are democratizing the payments industry from incumbents such as banks, acquirer networks, and so on.
  - Apps such as iTunes, Google Play, Amazon, eBay, and so on have direct and full financial access to millions of customers worldwide, and based on an OTT service with the utmost agility to bring innovations and convenience. As an example, iTunes has more than 400 million active credit cards worldwide.
  - Mobile device trends are moving into mobile payment and eSIM standards, where relationship is between the App developer or OEM and the final customer, most of the times without middlemen (like the role telcos have been playing for a long time).

- *Advertising*
  - Inventory is fast moving from web to apps. Currently, there are more than 2.5 million apps (IOS and Android) with 1000 new apps every day.
  - Average CPI is growing at 14% (2014–2019) and some markets, such as Germany, are well above the worldwide average CPI.
  - Indeed advertising is growing so much in mobile that ad blockers and similar techniques are appearing. Industry is responding with smart notifications, customized to each customer needs, and triggered based on a complete Business Intelligence model that measure behavior across dozens of customer KPIs on real time.

- *Media and content providers*
  - Apple iTunes, Google Play, smartphones, and tablets changed everything. In a few years, users were able to access anyplace, anytime their preferred content on any device. And that is a major change.

- Leading content providers, such as Disney, Dreamworks, and so on, had to adopt new formats, new services and new business models, sharing part of their revenues through new and premium distribution channels.
- IOS and Android stores established 30/70 rules. So publishing your content/app in these stores enjoyed a worldwide distribution though app developers have to share a big part of the cake.
- New models such as Spotify/Napster for music, Amazon kindle unlimited for books, or Netflix for movies/series changed how industry distributed and how customer consumed contents. Flat monthly subscriptions with access to all content were a completely different story. As these models were born OTT, these companies were able to micro-segment their customers' base, improving efficiency to levels never seen before. As an example, Netflix manages more than 70,000 different microsegments, offering personalized content to each of them.
- Other two great examples are Amazon opening retail stores in the United States to compete with the remaining 600 Barnes & Noble retail stores and Google offering Google Fiber (1 Gbps) and Google Project Fi (MVNO) in the United States to challenge established telcos in front of the customers to push for better prices, offers, and technology.

Therefore, all these changes in combination (app stores, smartphones, etc.) created a new ecosystem where adjacent industries needed to revisit their business models as new players emerged, sometimes more powerful than established companies. This multilayered value chain is on constant move. As another proof point we find Netflix that started succeeding as an OTT distributor, multidevice, and soon after they tested to produce their own series, "House of Cards," investing US$100 million. Industry (including telcos) soon shouted they were crazy, but just a few years after they demonstrated success in increasing customers, reducing churn, deleveraging content producers, and reinvesting company profits into new series. Three years after they have more than 100 owned productions and this is being part of their company DNA. Customers love it and some telcos are starting to follow this path.

9.4 TELCO'S ROLE AND CHALLENGES IN THE NEW PARADIGM

Traditional telcos with decades of legacy have been historically embracing little changes to their business model, such as SMS for mobile industry. Core business has been, for a long time, the core business model, the killer service. In the pre-smartphone era, SMS started to grown, resulting in a new type of revenue stream for telcos, though the most optimistic forecast arrived at a 15% revenue coming from SMS as the maximum revenues they would be able to make from this business. Then data started to grow and SMS growth was not as expected, so the forecast continued at 15%, now mixing SMS and data. But smartphone, 3G speed coverage, apps, and democratization of WiFi started a no return path to data usage growth, as we never saw it before in any other service. When in the 1990s we were talking about kb for speed and Mb for monthly volumes in fixed network, today we talk about hundreds of Mb for speed and GB for monthly volumes. In mobile we came from 9–14 kbps for GSM to 100 Mbps on 4G.

And most of telcos sooner or later understood this change and focused on data strategies, investing heavily in new generations of fixed line and mobile networks but, at the same time, looking for efficiencies and synergies through MNOs in country consolidation, network sharing strategies, and so on.

Though soon data monetization became one of the strategy topics in board agendas, most telcos did not react in time to another (parallel) phenomenon: OTT business using data and disrupting telco business and adjacent industries. Once customer was able to access data at relevant speeds, everywhere and at a reasonable cost, smartphones brought a new way to access contents, features, and services through apps.

OTTs were able to build multibillion dollar business from zero and in a few years. As a proof point of this huge change we can have a look at iTunes. iTunes started in 2006 with iPhone and zero revenues. In fiscal year 2014, iTunes[3] accounted for $18 billion revenues with 85 billion App Store downloads. Just in 8 years Apple built a truly new multibillion business from zero, revolutionizing adjacent industries in the journey.

OTTs and the new players enjoy a completely different way of doing business:

- They don't have legacies and don't need massive investments; they need to recover like network investment for telcos.
- Time to Market is about hours/days/weeks, not months/years.
- They have direct reach to customers so they don't need telcos or other players acting as distributors/middlemen.
- They can launch quick and fail and then launch a new product/service and succeed. Innovation and customer convenience is part of DNA.
- Their business is growing really fast, creating billion dollars in just a few years giving them cash to reinvest and move forward.
- They are truly global, not caring about borders.
- Decision making and company employees' average age make possible flat structures with real-time reactions (start-up-like operations).
- Recruiting the best and 24×7 motivation are part of these companies DNA.

Facing this change of paradigm, telcos reacted differently. In general, we could say that telcos from Southeast Asia such as SKT in Korea and NTT DoCoMo in Japan reacted quickly and today VAS on top of voice and data represents a big part of the revenues cake. Although how they arrived to this revenue mix is not 100% clear, probably country culture and the fact that they started to develop VAS heavily on the presmartphone era give clues about it. These telcos grew a new brand perception based on innovation and became quickly VAS strong players that allowed them to enter into the VAS/OTT game, creating their own app stores (such as SK Planet), creating new categories (such as SKT Joon 2, a smartwatch for kids), and giving value to their customers far ahead of their traditional voice, data, and VAS.

On the opposite end, western telcos, including U.S. ones, managed to react by creating special structures fully focused on digital and new VAS. Nevertheless, the main focus was still their core business and their real priority, so growth in the new area was always limited to a fraction of what was needed for the telcos to be considered as a big player in these new businesses, compared with the OTTs.

Indeed, many telco executive boards have been discussing for a long time whether the telcos should become a utility or take the transformational step into a digital telco able to face OTT challenges. Most of the times, the utility decision won (hands down) or by *de facto*. Some telcos bought some OTTs, such as social networks or web search engines, but the difference between their core business and these new businesses made it difficult to work together and sometimes the acquisition drove these new businesses to extinction, leaving more power to remaining OTTs in the game and becoming their competitors today.

9.5   BUT WHAT DO WE UNDERSTAND BY VAS TODAY?

VAS concept has evolved from fixed line voicemail to, for example, requesting taxi service through the mobile phone. And it is still evolving in its own nature in and out of the telco value chain, offering incredible changes in adjacent industries (look at Uber or AirBnB versus traditional taxi service and companies such as TripAdvisor.com to book hotel rooms).

The real thing is that customers, millennials, and customer needs trends are the ones defining or accepting new services and new VAS. And brand perception is quite important. It is not just about having a service. Many times brand, social presence, word of mouth, and so on are more important than the service itself. Companies that knew to manage all these coordinates were successful versus others, even if pushed by traditional telcos with vast presence in the customers' minds and pockets. Take messaging companies such as WhatsApp, Telegram, Viber, and Skype as examples. SMS is declining rapidly due to these new companies. And some of them, such as WhatsApp, have included voice calls in their service to more than 1 billion customers. Access to these customers is the key to try new services and change the business models and even industries in months or a few years. At the beginning, telcos argued that the quality of these services was not good enough, but they have been improving a lot to the point that in some countries, customer perception about quality is better for OTTs rather than traditional telco "carrier grade" voice services.

Customer needs are evolving quickly and adapting even faster to new (and more convenient) services. Indeed, this new kind of access to everything through Apps is changing traditional behaviors such as watching TV. PC/laptop is the most common video/TV watching platform for young people, and they do not see streaming and linear channels. Now that telcos have positioned well with great Pay TV offers, next generations are not their next customers?

Analyzing content's value chain, it seems clear to me that content is really the king. If we look at multicountry, multiplayer distribution, content benefits and revenues, customer engagement, and so on, we will find that content moves the full value chain, including that of the telco. Let us stop 1 min into this: Disney[4] bought Marvel Entertainment for $4.2 billion in 2009 (300 employees, +5000 characters), which was something around 10 times EBITDA at that time. Just the Avengers movie represented a box office[5] of $1.52 billion worldwide. All Marvel movies to date from acquisition by Disney hit $9 billion just in box office with a production budget of around $2 billion. And Disney really masters the art of merchandising. The future is all about content.

Nevertheless, let us get back to the more traditional telco VAS.

- *Financial services*

    Within the financial services vertical, we can find several products, but the most widespread ones are mobile insurance and carrier billing. Telcos in emerging countries, where the majority of population does not hold a bank account nor a credit card but hold a mobile line, are the ones that can leverage their well-developed carrier billing to offer different services to unbanked customers (Google Play Apps, access to Spotify, etc.). Carrier billing is a great asset telcos can offer to partners and it is a good opportunity to access to customer's wallet without a traditional banking scheme. Indeed even top brands such as Apple are now considering carrier billing to access certain levels of customers.

    Mobile payment is another growing VAS either for customers or for partners. Nevertheless, this is a more complex business: established traditional players such as banks and payments-acquirer networks (national and international) are very keen to protect their traditional business and market share, most of the time offering freemium fees for payments acquiring services if they are able to capture a merchant's bank account and activity. On top of it, this business is quite fragmented, that is, you really need to reach an agreement with several acquirers, banks, and POS OEMs to be able to offer a viable payments system to merchants. But where we can find a massive fragmentation is in retail where you may need to partner with thousands of retail companies, just in one country, to increase new payment scheme adoption (such as mobile payments), with different security requests and risk parameters and many times considering a short-term investment (development, hardware, etc.). Merchants and all players know that mobile payments is the future, offering more convenient services and will drive massive business, but this is long term as to allow all ecosystem players to be on the game in the short.

    Therefore, mobile payments is a constantly evolving business and I believe we will get to a point where everything will be OTT payment, with no notes, credit/debit cards, checks, and so on, used by customers, but probably this will take some time.

- *Security and Cloud*

  Maybe security so linked and connected to the core telco offering is the vertical that has evolved very similarly across telcos. In B2C parental control solutions, cloud storage, antivirus, and so on are the most advanced solutions, either in fixed line or mobile, with lots of similarities in all telcos. What we find here is that only 10–15% of customers really install these types of solutions, although 40–50% know they need one. Again, brand recognition and business development are key to start growing new categories and maybe telcos faced security B2C as a pure optional VAS instead of making it as part of the core offering. Only a few telcos made it part of the core offering, but success did not boost core results significantly.

  Finally, we also find another interesting product for B2C: smart home including security, energy efficiency, and appliances management, all in one. Many big telcos are adopting their own solutions (such as AT&T, DT, Telefonica, etc.) though it is not always an easy journey. Telcos need a powerful IoT platform and customers have to face upfront cost of no less than US $150–200. And then they need to pay a monthly fee. There are many DIY solutions and established security solutions in the open market and retailers that compete with telcos' solution in the customer's mind. And here we arrive at the brand perception issue again. Telcos need to change itself to allow customers think about telcos in a different way. This "carrier grade" new services need to be educated in customer's mind as to offer proven and trusted options.

  B2B is a completely different story. With hacker's activity increasing and technological and programming skills growing in many countries, companies have identified security as a core part of the offering. Indeed, data and security are even more than 40–50% of telco bills for large companies. Telcos understood this quite fast and got ready to offer security solutions partnering with top brands support.

- *Education*

  With some forecasts of hundreds of billions euros of revenues worldwide in 2020, digital education is one of the fastest growing markets for the next years. Nevertheless, there are several challenges to consider.
  - As of today, there is huge fragmentation in terms of where to find content (app stores, online, off line, etc.), customer's budget (customer's wallet is different according to the country), lots of competitors, and so on.
  - Different localizations are required as different regulations, national education systems, and methodologies are governing the current education system in several countries.
  - Education should be "universal" and accessible through mobiles, data, or fixed networks/laptops/PCs. Nevertheless, coverage and devices penetration is not yet universal, specifically in emerging countries where access to education is still under development.
  - Offers for each age group should be supported either by kids characters (such as Disney, Lego, etc.) or top schools/universities who that able to certify quality and provide curation and updated contents.

Although digital education is a huge opportunity, tackling it is not so easy, but several telcos are making progress offering mobile learning platforms and partnering with well-known universities or established content providers. The business model is then another story. With most of them considering education as a VAS and not part of the core, some use freemium models with in-App purchases and others use monthly flat fee with full access. But meanwhile customers can get lots of free educational contents in several apps (such as Apple University).

• *Entertainment*

After an initial period where telcos were trying to do their own service (e.g. music), it became clear to them that premium partnership with trusted and established brands is the most successful model. Top brands such as Napster, Spotify, EA Games, and so on are partnering with telcos to offer special bounties for customers.

In the case of video, most of top telco players have developed their own IPTV service, offering multiplatform–multidevice linear channels, VoD, and live premium content (mainly sports). Nevertheless, new generations (millennials, etc.) are changing how they watch TV, video, or live streaming. Telcos should adapt quickly to this new trend to stay on the customer's top of mind. However, it is not going to be easy as offering such a wide content requires major investments. For instance, Sky[6] agreed in 2015 to pay £4.2 billion for 2016–2019 premier league rights in the United Kingdom. This is about £1.4 billion per year for a company hitting £7.8 billion revenues in the United Kingdom. Programming costs are increasing (at least live sports, which today is the main differentiator from potential OTT competitors) and not all companies are able to face this level of investments to maximize EBITDA with such a high and growing costs structure. It is not a simple equation, but live content may keep the telcos alive in face of OTT challengers.

• *Advertising*

Advertising has evolved a lot in the recent years. Most of budget is moving from web and even TV to mobile. With mobile video consumption increasing rapidly and new generations trend to move to mobile, advertisers need to position themselves in this arena. But the customer is no longer open to receive any type of ad. Indeed many people are adopting ad blockers and similar tools so as not be spammed with undesired content.

On the other hand, the customer's knowledge is exponentially increasing through apps use, smartphones, and always on "way of life." Just think about driving your car with Google notifying you that it will take 30 min to work with current traffic conditions, without any request from the user. Business Intelligence is key and publishers are very conscious of this, because having the right one and building the right triggers, publishers might be able to adopt advertising to every single person, increasing conversion ratio incredibly.

Another very good example is Netflix, with a recommendator engine that triggers customized notifications for every customers' segment, but they have more than 70,000 segments!

Telcos are heavily investing in BI as they know this is and will be key for engaging customers and one of the most valued services for publishers and content providers.

• *Health*

We can find several health initiatives, some of them very interesting, such as health portals trying to democratize health in developing countries, where many people do not have access to any health support because they are really far from health centers or they do not have enough money to afford it. These portals are making health support available through any device and any connection, so many people can access this service.

There are other interesting initiatives around online surgeries (Google glasses) or chronic diseases, but they are not massive and linked to B2B, so fragmentation, localization, and regulation come to the table again.

Another issue that has achieved major progress is devices. Health parameters have been started to be tracked by some smart watches and other devices, but the quality is not yet at the grade level requested by national health authorities nor accurate enough as to produce a relevant customer experience, so they are not massively used. Nevertheless, technology is making fast steps forward and I am sure we will see a dramatic change on health ecosystem, monitoring, and tracking in the next few years.

9.6 SO WHAT'S THE FUTURE FOR VAS AND, THUS, FOR TELCOS?

Telcos have done well to date—facing several crisis, in-country consolidations, Fixed-Mobile Convergence, and so on, in several countries, maintaining or slightly declining revenues and offering more digital products and VAS to customers. Nevertheless, the telco industry is facing a massive turnaround and I see some telcos more prepared for this than others.

VAS is no longer what we understand by traditional VAS concept, but it is the *new core service*. Customers expect to find connectivity *de facto*, either WiFi, mobile data, or tethering. The core service will no longer be data. Data connection is a basic, "as value in the war." Customer is choosing new services based on Apps and features, so services for customers are video streaming, booking a taxi, buying holidays, or making business. We need to understand this and telcos need to partner with the top brands in each category to present special offers for their customers. So rivalry is no longer for customers, but to make the best partnership and present the best offer to customers.

Telcos are facing a new dilemma: They can evolve slowly using Business Intelligence, data, and networks as triggers of new revenue models or transform themselves into hybrid players (half network and half OTT) to compete face to face with major OTT players to maintain the power of the network. This second option is not obvious as network is territory based and OTT does not understand country borders. Some small OTT-like initiatives coming from telcos are gaining traction, although they do not represent a major transformational bet.

The transformational option would require telcos to build on top of the brand (or create a new brand), educating customer, and adopting start-up techniques to really being able to compete with OTTs. To date, some telcos tried it with lack of results, normally acquiring and "killing" an OTT business, by being lost in the telco "traditional legacy and culture ocean."

Nevertheless, there are plenty of opportunities out there, especially in emerging countries where unbanked ratio and access to Internet do not cover all the market today.

So the question is, will telcos "take the blue or the red pill"?

ACRONYMS

B2C	Business to Consumer
B2B	Business to Business
CPI	Cost Per Install
DECT	Digital Enhanced Communications Telephone
DIY	Do It Yourself
EBITDA	Earnings Before Interest Taxes Depreciation and Amortization
eSIM	electronic SIM
FMC	Fixed Mobile Convergence
IoT	Internet of Things
IPTV	Internet Protocol TV
KPI	Key Performance Indicator
MVNO	Mobile Virtual Network Operator
OEM	Original Equipment Manufacturer
OS	Operating System
OTT	Over-the-Top
PC	Personal Computer
SMS	Short Message Service
TMT	Telecommunications Media and Technology
VAS	Value-Added Services
VoD	Video on Demand

NOTES

1. I want to thank Antonio for giving me the opportunity to contribute to this amazing book as I believe this topic is one of the most relevant ones to talk through when analyzing our industry and trying to predict what's next.

2. Seven years into the mobile revolution: Content is King . . . Again. Yahoo! Developer Network, https://yahoodevelopers.tumblr.com/post/127636051988/seven-years-into-the-mobile-revolution-content-is (August 26, 2015).

3. iTunes by The Numbers: $4.6B revenue for Q4, $18B total revenue in 2014, 85B App Store Downloads. MacRumors, //www.macrumors.com/2014/10/20/itunes-by-the-numbers-q4/ (October 20, 2014).

4. 2010 Annual Report Disney. The Walt Disney Company, https://ditm-twdc-us.storage.googleapis.com/2015/10/2010-Annual-Report.pdf (January 1, 2011).

5. Box office history for marvel cinematic universe movies. The Numbers, //www.the-numbers.com/movies/franchise/Marvel-Cinematic-Universe#tab=summary (last accessed April 14, 2016).

6. 2015 Annual Report Sky (year ended June 2015). Sky, 2015, https://corporate.sky.com/documents/annual-report-2015/annual-report-spreads-2015.pdf

Chapter 10

Mobile Virtual Network Operators/Second Brands

Jaime Bustillo

Why MVNOs? The reason for MVNO to exist is to increase the number of players in the mobile services ecosystem. They are not needed for any technical reason, as they do not own or operate an exclusive piece of technology or get a license to use scarce resources.

Their only reason to exist is rooted into two main facts that set very high entry barriers for new companies to enter into the Mobile Network Operators (MNO) arena and effectively limit the number of players who may operate in a given area:

1. *Mobile Network Operators need frequencies to operate on:* The spectrum is limited, and the behavior of different portions of the spectrum is different, making it very difficult to allocate enough frequencies with the same competitive advantages to many different MNOs in the same area. While there are extreme cases, three MNOs is the optimum number to spread the spectrum into if we want them to have similar competitive situation from radio propagation and capacity point of view. In many countries this number is only two, or may be four. In case of two, the situation is considered a "Duopoly," with not enough competition. In the case of four, it is not easy to divide the existing spectrum into four equivalent pieces while ensuring none of the four is disfavored from the investments or quality of service point of view.

2. The investments required to build a new MNO network in a country are extremely high, in terms of both capital expenses (acquiring, building, optimizing a new radio network) and operational expenses (site leases, preventive and corrective site maintenance, operation of the network, transmission, and transport costs). These very high costs combined with the big price reductions that were applied in the last 20 years to end customers make it difficult to build and operate a fourth network in a given country, and still be able to continue investing in network modernization.

Digital Services in the 21st Century: A Strategic and Business Perspective, First Edition.
Antonio Sánchez and Belén Carro.

There are other considerations, such as finding space for new antennas at rooftops or the visual impact of towers in rural areas that have also started a trend in the last few years of sharing part of the infrastructure even for established MNOs.

The MVNO ecosystem is composed of two groups of companies:

- The many different flavors of MVNOs, who at the end only have in common the fact that they sell mobile services to end customers using their own brand. These are Full MVNO, Service Providers and Branded Resellers, as the main kinds of MVNO.
- And a small group of other players who support the MVNOs in some of their functions, and do not serve directly end customers. These are known as MVNAs (Mobile Virtual Network Operator Aggregator), MVNEs (Mobile Virtual Network Operator Enablers), and Technology Enablers.

We will explain the ecosystem in the next few sections of this chapter.

10.1   FROM OLIGOPOLY TO MARKETPLACE

As we explained earlier, the characteristics of the infrastructures and costs needed to build and operate a mobile network make it very difficult to have more than three or four players in the same geographic area. This in the past was an unacceptable limitation to markets.

This oligopolistic situation limited competition in the market, setting a de facto lower limit for the prices to end customers, and favored the existence of giant companies that are vertically integrated and inefficient as it usually comes from this vertical integration of a business. Many products and services were not developed and delivered to customers, if not for other reasons, as they were reprioritized in the control processes of these giant vertical integrated companies. On the other hand, there were no incentives to be efficient in many areas of the oligopolies, and not enough competition to reduce the end customer prices.

The envisioned solution in some more advanced countries on these aspects was to split production from distribution and sales of mobile services, in a way that would allow many players to distribute and sell the mobile services to end customers. This would increase the competition, reduce prices, and allow the entrance of more specialized companies that could explore the whole span of services possible. At the same time, this will force the former vertically integrated MNOs to be more efficient on their distribution and sales functions.

An example of this increased efficiency forced by the MVNOs in many countries is the reduction or complete elimination of the terminal subsidy that came from the old historic PTTs and the need to deliver the fixed line phone in most cases with the phone line. The terminal subsidy was hiding the cost of the terminals behind the very expensive usage tariffs that were keeping the usage of services per customer from growing These artificially inflated tariffs, were paying a different section of the mobile ecosystem: the terminal manufacturing and distribution. The MVNOs in most cases do not have the financial muscle or the logistic infrastructure to be able to distribute the terminals and much less to subsidize them, and cleared the use tariffs of this additional load, making transparent the prices for each component of the mobile ecosystem and reducing the end customer prices.

With the introduction of MVNOs, and their ecosystem as we will explain later, the market can grow to multiply its dimension by a factor of 10 normally, with a figure of 40 MVNO being normal in markets that have been successfully opened to MVNO competition. This moves the mobile services market from oligopoly to marketplace if it is well executed.

The problem when trying to create this marketplace is that it requires first to create other marketplace: the wholesale services marketplace, where the MVNO will buy services and traffic from the MNOs. This marketplace is by definition oligopolistic if you do not take special measurements, restricting the MVNOs from starting up and growing. This always requires regulation to a big or small extent.

You can regulate the wholesale prices: This is a starting point, but will not deliver good results in the long term, as these markets are normally much more dynamic than the regulation that should be setting up prices with the same dynamism of the market. What usually works best is to force the players to reach agreements, regulating disagreements only in case no agreement is reached between two individual parties.

With a successful regulation in some years (5–10), the MVNOs market share in the total mobile services market could be around 15–20% of the market.

10.2 MVNO ECOSYSTEM: END CUSTOMER FACING OR MVNOs

This part of the ecosystem is composed of the different kinds of MVNOs that may appear in the market. They will differentiate on the portion of the value chain that they are performing, and this will create differences in company size, skills required, portion of the total margin controlled, and so on.

We can see the different kinds of MVNOs mapped into the value chain components in Figure 10.1. These different components an MVNO may capture from the value chain can change a lot, and are evolving continuously in the last few years. We will try to give here a simple classification of the MVNOs.

10.2.1 Full MVNO

This is a kind of MVNOs that includes almost all the components into the value chain, only excluding the radio and transport network. These operators are sometimes called Thick MVNO.

These operators need a license to operate, and will need to be allocated different numbering components by the national authorities to operate, as they appear to the rest of the world as an independent operator from the signaling point of view, with no difference to any MNO or any other Full MVNO.

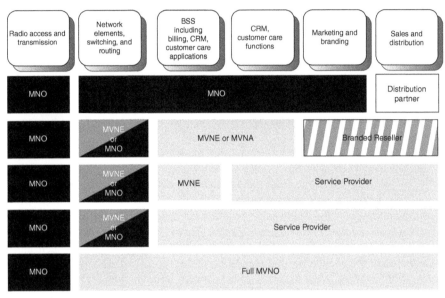

Figure 10.1   Different value chain segments define MVNO kinds.

These numbering components are as follows:

- *MCC: Mobile Country Code:* A three-digit number that identifies the country where the MVNO is operating.
- *MNC: Mobile Country Code:* A two- or three-digit number that identifies the MVNO inside the country of operation.
- *ICCid ranges: Integrated Circuit Card ID:* The MVNO will need to be allocated a range of ICCid numbers to be able to produce their own SIM cards.
- *MSISDN ranges:* The MVNO will need a range of MSISDN numbers allocated by the national authorities inside the country's phone numbering plan.
- *NRN: Network Routing Number:* This number will only be needed if the country has implemented number portability, and will be used to route the terminated calls to the network where the particular MSISDN has been ported.

The number of Full MVNOs that can be allocated in a single country is not limited. In practice, operating a full MVNO will only be a good option if the number of subscribers forecasted is above 200,000.[1] If the number is lower, with high probability the regulatory obligations and consequent costs will be bigger than the added margin earned.

There has been a special case of Full MVNOs appearing in the last few years that are the international Full MVNO, who use the same network and BSS resources to operate as Full MVNO in different countries. It has some cost advantages and allows the operator to provide the same products in different countries at a lower cost.

An MNO operator can be an MVNO in a different country to complete its footprint and to use its brand and technical assets in a broader market.

10.2.2 Service Provider

While there could be a variety of ways to implement a Service Provider, the key difference with the Full MVNOs is that Service Providers do not have the network components needed to operate the service, in terms of both radio and core network assets.

A Service Provider will, in general, have its own BSS/billing systems that will allow it to operate with its own set of products and services. The BSS/Billing could be its own infrastructure or provided by an MVNE.

A Service Provider will need to have an operator license, and will require the suballocation of some numbering assets, not all that were required by a Full MVNO:

- *ICCid subranges: Integrated Circuit Card ID:* The MVNO will need to be allocated subranges out of the ICCid numbers of its host MNO or Full MVNO to be able to produce its own SIM cards.
- *MSISDN subranges:* The MVNO will need a subrange of MSISDN numbers allocated by the national authorities from its host MNO or Full MVNO MSISDN numbering ranges.
- *NRN: Network Routing Number:* This number will only be needed if the country has implemented number portability, and will be used to route the terminated calls to the network where the particular MSISDN has been ported.

The number of Service Provider MVNOs that can be allocated in a single country is not limited. In practice, to operate a Service Provider MVNO will only be a good option if the number of subscribers forecasted is above 10,000. If the number is lower, with high probability the regulatory obligations and consequent costs will be bigger than the added margin earned.

10.2.3 Branded Reseller

The minimum numbers stated above for Full MVNO and Service Providers will leave many other companies who would like to offer Mobile Services out of the market, such as medium-size brands or local cable TV companies based on cities of around 10,000 habitants.

For those companies, and number of customers below 10,000, the figure of a Branded Reseller can be applied. A Branded Reseller will in general only provide Branding and Marketing, and Sales and Distribution functions of the value Chain. They can also provide customer service, but only if these small operations have it already deployed for the main activity of the company.

A Branded Reseller in general does not need to have a license and it can operate under the umbrella of his MVNE or MVNA license, who would be responsible for all regulatory and legal aspects of the Mobile Services provision, while the Branded Reseller would only be responsible for the customer contact and management.

But we can find Branded Resellers with or without a license, and served by an MVNE or a MVNA, depending on the actual possibilities offered into the particular country.

The Branded Reseller figure is located between a Service Provider and a Distributor of one of the big MNOs, the only difference is the brand used, and there may be some differences into the product offered that may be more tailored for a niche market instead of being the general product offering of an MNO.

A Branded Reseller could be very small, down to 1000 or a few hundreds of customers, depending on the country or the company behind the offering. If it is a standalone operation, it should not be below a few thousand customers.

10.3 MVNO ECOSYSTEM: TECHNOLOGY ENABLERS, MVNE, AND MVNA

This part of the ecosystem is composed by the different companies created and developed along with the MVNO market itself, as it was growing and diversifying, and specialized solutions were needed.

These companies are in general not end customer facing, but B2B acting between MNOs and the MVNOs.

10.3.1 Technology Enablers

From very early on, just after the first wave of Full MVNOs started by the Fixed Telephony companies went live, it was evident that for most of the next wave of players, technologies and products developed for giant vertically integrated operators were not suitable for the new generations of MVNO and they were far too expensive to be used.

The different MVNOs were too small for the big Telecom Vendors, they were too small too for the big IT firms to deploy and integrate their products, and both will require a large amount of resources at the MVNO to specify, contract, accept, and maintain the products from those companies.

A small group of companies appeared, navigating on the waves of Virtualization and Software Defined Networks, that were making it efficient to add many network and IT functions in a reduced set of servers, and at the same time, to share these functions on different instances for different MVNO customers. With this as the driver, these technology enablers enabled a different set of Full MVNOs and after that, Service Providers, who are technology and technology knowledge lean, as they do not need to own and have all the knowledge anymore to operate what used to be very specialized and complex networks.

Today, there are some global companies, usually start-ups created to perform this function by former executives of Telecom Companies, and in some cases, attempts from some of the big vendors to enter this market, where they can be only successful by enabling the biggest MVNOs, due to their very heavy company architectures.

A handful of local new entrants usually attempt to enter this market with the opening of every new market to MVNOs, making sure this part of the ecosystem will continue to be alive and in good health.

10.3.2 MVNE (Mobile Virtual Network Enabler)

An MVNE is a company that has a contract with one or more MNOs to operate as a Full MVNO, but his activity is not dedicated to end customers, but to act as a host for Service Providers and in some cases Branded Resellers, providing them with all MNO relations, the needed network elements, and in some cases CRM or customer care functions.

They are a very important part of the ecosystem, as they enable Service Providers to operate without creating a big workload to the MNO on serving them, and enabling simpler companies to operate in this market focusing themselves in marketing and sales activities, but not caring much about the technologies needed to operate the services.

The original MVNE grew up in the markets that were first opening up to MVNOs, such as Sweden and the Netherlands. We can see in Figure 10.1 how these MVNE are positioned in the value chain, occupying one or more spaces depending on the needs of their MVNO customers.

An MVNE in general will enable more than 10,000 MVNO forecasted customers, Service Providers, or Full MVNO to provide service. If the MVNO customer is smaller, another player will be required that we will discuss in the next section: the MVNA.

The MVNEs are also much faster in setting up a new MVNO and doing it much cheaper, with the revenue guaranties and the liabilities softer than those of the MNOs if you do a direct connection.

10.3.3 MVNA (Mobile Virtual Network Aggregator)

The function of MVNA is to aggregate many smaller MVNO (Branded Resellers normally) with a very standardized offer and functionality to reduce the development and operation costs, to get enough volume to connect to an MNO or an MVNE,

The MVNA will have its own OSS/BSS and CRM/customer service functions and applications to serve his Branded Reseller customers. An MVNA provides a good turnkey solution to extend the feasibility of selling and operating mobile services to companies of the smaller tiers. In some countries where the markets are already very saturated, the MVNAs are located at the top of the list of customer growth between MNOs and MVNOs

The MVNAs would serve companies with a forecast of a few thousand customers, aggregating them to 100,000 or a few hundred thousand to be connected to an MVNE or MNO directly. While an MVNE will normally try to make room for any request it receives, an MVNA would normally specialize in a sector where a unified proposition may work without further adaptation for each customer.

ACRONYMS

ICCid   Integrated Circuit Card ID
MCC   Mobile Country Code
MNC   Mobile Network Code
MNO   Mobile network operator
MSISDN   Mobile Station International Subscriber Directory Number
MVNA   Mobile Virtual Network Aggregator
MVNE   Mobile Virtual Network Enabler
MVNO   Mobile Virtual Network Operator
NRN   Network Routing Number

NOTE

1. These numbers are practical numbers taken from the industry. They may change depending on many different variables.

Chapter 11

Digital Home

11.1 INTRODUCTION TO HOME AUTOMATION

Home automation is a technological field that is currently attracting many futurist ideas but is also suffering the consequences of the high innovation velocity and the lack of consensus in standardization. Today there is a huge variety of devices that can be found in the market related to home automation: bulbs, switches, door and window locks, thermostats, home appliances—most of them controlled by applications installed in smart handheld devices such as smartphones or tablets. Also, many wireless connectivity technologies are being employed by the devices for control purposes, including ZigBee,[1] Z-Wave,[2] or Bluetooth 4.0+.[3] Individually, the devices work properly and are generally easy to set up and use, but unless they belong to the same ecosystem or manufacturer, different applications and even different physical elements, such as bridges, will be needed for the system to operate. New global solutions are appearing to cope with this issue, but at this time interoperability problems are one of the reasons for the slow implantation of home automation. High prices of home automation global solutions along with the user unwillingness to undertake inconvenient works at home also contribute not to expand the aforementioned sector. Manufacturers have tried to solve this issue by putting into the market "plug-and-play," economic and scalable solutions: the home automation kits, such as energy efficiency kits, security kits, water leakage kits, and so on.

The current interest of home automation is perhaps driven by two main targets: security and energy efficiency. Security includes personal security (e.g., antitheft by using IP video cameras) and home security (e.g., gas or water leakage notifications), and energy efficiency includes electrical and fuel consumption savings, mainly in heating and air-conditioning systems but also for home appliances electrical consumption control.

Digital Services in the 21st Century: A Strategic and Business Perspective, First Edition.
Antonio Sánchez and Belén Carro.

11.1.1 Home Automation Basic Elements

A home automation system is also called domotics system. It consists of several basic elements and for simplicity we can cite sensors, actuators, control system, and transmitters.

- The control system manages the domotics system in centralized infrastructures. It comprises the system intelligence and usually has the user interfaces needed to transmit and receive the information, also allowing for monitoring and controlling the different subsystems, providing multiple functionalities such as stocks control, preventive maintenance or external services requests for example, security, repairs, or supplies.

- The transmitters receive the user directions and send them to the control system. Push buttons, switches, touch displays, remote controls, and IP gateways are examples of transmitters.

- Sensors are devices that collect environmental information and send it to the control system to act consequently. For example, an anemometer measures the wind speed in order to determine whether the awning is to be automatically closed. Usually they use batteries, so they are not connected to electrical power, which implies flexibility. Also, some sensors may directly communicate with actuators or even share the same device. Among the most common sensors we can find are room thermostats that control the room temperature; inside temperature sensors that just measure the room temperature; outside temperature sensors that may optimize the heating system performance; temperature probes for heating management that allow the correct control of different types of electric heating, for example, limiter probes for radiating floor; humidity probes that detect possible water leaks in the kitchen, bathroom, and so on; gas leak detectors, especially in the kitchen; smoke and/or fire detectors that detect fire outbreaks; radiofrequency detectors that receive medical alerts from a radiofrequency portable transmitter; volumetric and perimetric presence sensors that detect house intrusions; environmental lighting sensors to adapt artificial lighting; infrared receivers, and so on. Besides, several sensors may be used jointly for a specific task, for example, temperature and lighting: If it is hot and there is a lot of sunlight, then lower the blinds.

- The actuators are elements used by the system to modify the state of some equipment: They execute the commands sent by the system intelligence. The most common actuators include electromagnetic relays, solenoids for cutting off the water and gas supply, sirens for alarms and zoning valves for hot water heating. For example, if a smoke sensor detects fire, it will alert the control system, which will establish the programmed telephone communication and will act over the house gas shut off valve and over the siren. Both sensors and actuators may be integrated in a unique device.

The control system may be wired, wireless, or mixed. Wired stations use cable to communicate with every sensor and actuator. Wireless sensors powered by batteries transmit through the radio spectrum the event information to the wireless station. Mixed stations include both wired and wireless communications.

The network topology establishes how the different devices in a domotics system are interconnected. Common topologies are star, bus, ring, and mesh. A star topology is used in centralized systems where all the information goes through the controller via point-to-point connections. In case of failure of either a station or the transmission media that connects it to the central node or controller, the station will be isolated from the rest of the network. The maximum traffic rate is limited by the controller congestion. In a bus topology every station is connected to a single communication channel called bus via multipoint connection, and information is broadcasted to reach all stations. Any station may transmit information through the bus employing an access algorithm to avoid simultaneous transmission and therefore datagram collisions. When a station fails it is isolated from the networks but the bus is not affected. In a ring topology, every device is connected to another two, and a mesh topology implies an all-to-all connection.

11.2 EVOLUTION TO DIGITAL HOME

11.2.1 Definition of Digital Home

Digital home is a term with a wider meaning than home automation, since it incorporates ICTs (Information and Communications Technology) to control and program every home system inside and outside the house, through data networks, such as Internet, and using an appropriate interface. To become a digital home, an automated home needs to add the convergence of communications, informatics, and entertainment by using broadband networks. Another difference between home automation and digital home is the time span: While home automation or domotics exists since several decades and has undergone a slow evolution, digital home is relatively new and is on the upswing with an exciting near future.

Digital home is a wider concept that includes domotics, which is being relegated to just automation issues, and that adds multimedia content distribution in the home: video, audio, and data.

Different visions or concepts for defining the digital home have been around before. Telecommunications operators (telco) see the digital home as a (digital) connected home. Manufacturers prefer an automated home, that is, a home owning a series of devices allowing the control of several functions within a house. Domotics is an elder term that was formerly used instead of home automation, formed from latin *domus* plus informatics. In the same way, inmotics refers to buildings.

Some equivalent terms for digital home are smart home and connected home. Other terms are used along with digital home but they are not synonyms: green home and sustainable home refer to the energy efficiency aspects of a digital home.

Recent disruption in the digital home sector by giants such as Apple,[4] Google,[5] or Samsung[6] has pushed the market forward and has stimulated the interest of the technology for the sector. Other factors that have contributed to its current popularity can be cited: (a) smart devices, such as High Definition/Ultra-High Definition (HD/UHD) smart TV, smartphones, and tablets; (b) massive use of wireless technologies at home—WiFi,[7] Bluetooth,[8] cellular radio, and the devices that make use of them—stereo, home cinema, speakers, wearables, and so on; (c) multimedia communication technologies among devices, such as DLNA (Digital Living Network Alliance)[9] and UPnP (Universal Plug and Play)[10]; (d) advanced multimedia compression technologies—new video and audio codecs such as H.265[11]; (e) ultra-wideband access technologies such as Fiber-to-the-Home (FTTH)[12] and Long-Term Evolution (LTE)[13]/+; (f) energy control and saving systems—smart meters, individual heating consumption control, and so on.

In the near future, technologies, such as artificial intelligence that will allow the system to act consequently to what it has learnt from the user's needs and preferences, have a long way to upgrade the digital home capabilities for making the user's life easier.

There is a trend toward the integration of the digital home into the smart city, along with the smart grids. This integration is due to elements such as the smart meters, the electric car, and the decentralized electricity generation (e.g., photovoltaic panels), located in the smart home but also integrated into the smart grid, the intelligent electric network that includes public lighting, network automation, and bidirectional information transmission, among others.

The main current societal objectives related to dwelling are edifications with environmental integration and sustainable energy, and universal accessibility. The digital home can be defined[14] as a dwelling that, through technologically integrated systems and equipment, is able to offer to the inhabitants services that

- enable home management and maintenance,
- increase security,
- increase comfort,
- improve telecommunications,
- save energy, costs, and time, and
- offer new ways of entertainment, leisure, and other services inside the house and its environment.

Therefore, the digital home is offering services, thanks to the Information and Communications Technologies (ICTs) in the following areas:

- Communications
- Energy efficiency (diversification and energy saving)
- Security
- Environmental control
- Multimedia content interactive access (leisure, remote working, remote learning, and so on)
- Entertainment

Some of these functionalities are related techniques applied in edification (building orientation, insulation, etc.) but can also be maximized with digital home technologies, thanks to automation, technical management of energy and security, and so on. Therefore, an automated home would need convergence of communications, informatics, and entertainment, using the broadband networks, to become a digital home.

11.2.2 Services in the Digital Home

Starting with the communications service, it provides the transport of information (voice, data, or images/video) between the user and the different devices or services, or among devices that belong to the digital home. The communications services include basic telephony, broadband Internet access, home area network, IP telephony, and video telephony.

The digital home must be defined to allow an intelligent management of air conditioning and lighting, as well as the rest of the electric loads in the house. It includes management of electric devices, appliances, irrigation, water and main electric circuits, consumption monitoring, consumption control, and lighting control.

The security service allows the local or remote control of any area of the house and of any incidence related to the home, possessions, or people security, such as house intrusion, water leaks, or emergency management. The events are communicated with alarm notifications to the own user or to a Service Provider. The service sequence incorporates detection, notification, and operation, if needed. It includes smoke and/or fire detection alarm, gas detection alarm, flooding detection alarm in wet areas, intrusion alarm, SOS panic alarm, access control with video intercom and with proximity cards, video surveillance, and remote security (alarm reception central).

The environmental control service allows an integrated control of the different systems that employ the general services inside the dwelling, stimulating the achievement of the best:

a. Flexibility, being able to incorporate new services in the future, and to redistribute in the present, maintaining the existing services level

b. Economy, with an efficient use and management of energy: important savings in exploitation costs, maintenance, and structures simplification

c. Heterogeneous data integration, including management, control, and maintenance of every service and system inside the building and its infrastructures

d. Inhabitants comfort and security

e. Efficient communication for its operation and maintenance, with full automation of the activity.

The environment control service includes presence simulation, remote monitoring, remote control, awnings and blinds control, environment creation, air conditioning and temperature control, and remote maintenance and diagnosis.

The interactive access to multimedia content service incorporates contents, such as text files, documents, images, web pages, graphics, and audio, used to provide and communicate information. They represent data, information, and entertainment provided by several services to users at home, which can be delivered electronically or on physical media (CD, DVD, and so on). This service includes basic remote assistance, videoconference, remote working, and remote education.

By the leisure and entertainment services, people are able to enjoy their spare-time passively or interactively, through multimedia content that can be accessed from a player or display equipment. This content may be placed at home or received from external sources, through a broadband telecommunications infrastructure. The aim is to endow these services with the needed intelligence so that from the information and functionality provided by the multimedia digital devices and the individual social behavior, they may make decisions and envisage the user's needs to help with his or her daily tasks. The leisure and entertainment service includes radio broadcasting (AM, FM, DAB), Digital Video Broadcasting-Terrestrial (DVB-T), satellite and cable television, Video on Demand (VOD), multiroom multimedia distribution, IP television, online music, and online gaming.

11.2.3   Digital Home Infrastructure

Common telecommunications infrastructure[15] allows ICT to enter the home and provide both logical and physical support for providing the above mentioned digital home services. They include a broadband network access and a local network. Every digital home equipment must be interconnected to allow its management and control. The type of architecture deployed specifies the logical way in which different control and/or communications nodes, elements, and devices interact inside the house. When the network devices depend on a central system that regulates the communications and the control messages, we have a centralized architecture. On the other hand, in a distributed architecture the network elements handle their own process and control capacity, being able to establish communications with the rest of the elements by themselves. Finally, mixed architectures consist of centralized and distributed elements.

Communications are the essential elements that enable the new control services inside and outside the house. Traditionally, two different networks have been defined locally: a communications network also called Home Area Network (HAN) and a management, control, and security network, both interconnected by a residential gateway.

The HAN is deployed inside the house usually with structured cabling, and includes a data network and a multimedia network.

- The data network covers the communication services and is used for interconnection of computers, printers and other informatics equipment, Internet connection, and so on, and it employs common standards such as telephone wire, Ethernet,[16] WiFi, PowerLine Communications (PLC[17]), Digital Enhanced Cordless Telecommunications (DECT[18]); also connections between devices such as Bluetooth or Universal Serial Bus (USB[19]) can be considered as part of the data network. The multimedia network is used to capture, store, distribute, and play audio signals, images, and video and TV signals inside the house; it is broadband and employs common standards such as coaxial antenna network, PLC, wireless LAN, and so on. Currently, structured cabling networks are being replaced or complemented with radiofrequency networks, mainly WiFi and others such as small cells (femtocells, etc.).

- The management, control, and security network may be borne by other transmission media rather than cable, and it employs a variety of specialized standards as well as proprietary protocols. It includes an alarm and security network, which allows the connection of different alarms (fire, smoke, water, gas, electric supply fail, telephone line fail, etc.), intrusion alarms, video surveillance cameras, personal alarms, and a control network, also named domotics network, which interconnects sensors and actuators in control systems: lighting, air conditioning, comfort (blind, awning, or irrigation actuation), appliances, supplies, presence, and so on. It requires low bandwidth but high reliability, especially for security purposes.

However, at the moment of writing, and due to the rapid advance of technology, there is a tendency toward the disappearance of the home gateway, or else the integration of its functionalities inside the Internet broadband router, as well as to the disappearance and/or coexistence of cable networks with wireless networks for in-home communications.

For bearing any type of network a physical infrastructure is needed, to allow the information exchange and the needed communications among the different elements and devices, the devices and the users, the users and the digital home, or the digital home and the outside world. Two different bearers exist: the ones that need physical infrastructure deployment, and the ones that do not or that benefit the existing infrastructure.

- The first group includes cabling, optic fiber, and cable pipe systems. The majority of the current cabling bearers employ copper or aluminum conductors. There are two types of bearers: metallic wire pair and coaxial wire.
  - The metallic wire pair consists of two isolated conductor wires that may form multiple pair cables, twisted or untwisted, shielded twisted pair (STP, offering a better protection against electromagnetic interferences at the expense of a larger space needed and also more costly) or unshielded twisted pair (UTP). They may transport voice, data, and direct current powering.
  - A coaxial cable is formed by two concentric conductors separated by an isolating material with low losses. It is broadband and allows the transport of video and data signals at a high data rate. Generally, two kinds of coaxial cable are used: baseband, typically with 100 MHz bandwidth and 50 ohm impedance, and broadband, with a usual bandwidth of 400 MHz and 75 ohm impedance, employed for cable TV. However, some coaxial cables exceed the gigahertz bandwidth barrier.

Regarding fiber optic, it is composed of a silica core surrounded by a concentric silica layer called cladding, both surrounded by a plastic protective cover. Since the refraction index of the core is higher than the refraction index of the cladding, the fiber optic behaves like a waveguide. Two types of fiber optics are employed: monomode (thinner core and higher capacity) and multimode (thicker core and easier for interconnect). Minimum attenuation windows are commonly used for minimizing transmission losses. The fiber optic has several advantages, including reliability on data transfer, electromagnetic and radiofrequency interference immunity (the fiber optic may share the same duct with the electrical installation though independent ducting is recommended when possible), very high transmission speed, and high security for transmitting data. Finally, the cable pipe systems ensure the support, tied, fixing, and protection of cabling. They are installed inside the pipes (with embedded or shallow installation) or in protecting ducts and baseboards (shallow installation); they are easy to install and do not harm thermal or acoustic inolations in a wall, being a flexible installation.

- Wireless connection is the main type of bearer without the need of a physical infrastructure. It uses radio spectrum to transmit information, employing infrared or radio frequency technologies.
  - The infrared technology is widely used at home in remote controls for audio and video equipment, but its main inconvenience is its directivity that requires a line-of-sight transmission to avoid the communications channel interruption.
  - On the other hand, radio frequency offers high flexibility and ease of installation, however with the inconvenience of a high sensitivity to interference, which makes its efficiency depend on many external factors such as the type of materials used in the building construction.
- Finally, electrical wiring communication, also known as Power-Line Communications (PLC) is a technology that uses the existing electrical wiring in the premises for communication. It is easy to install, avoiding the need for users to install new wires inside the house, and is low cost for adapting existing buildings. Its main inconvenience is its susceptibility to the electrical noise inherent to the electrical wiring network, which is being overcome by new PLC systems.

11.3 HOME AUTOMATION: CONTROL NETWORK

11.3.1 Introduction

In the early times of domotics, automation of domestic equipment was limited to controlling its electric power, which constitutes a very easy to install management system but with low technological appeal since there was no efficient way of communication among devices. Domotics implies the incorporation of a series of systems that allow the efficient control and automation of the equipment and premises found inside the house. We can define the communication domotics protocol, also called control network, as the procedure employed by the domotics system to communicate with every device in order to control them. There is a variety of protocols, some of which are specifically designed for domotics. The protocols are classified into

- open standard or free to use,
- licensed standard that requires the following of license restrictions, and
- or proprietary for exclusive use of the manufacturer.

Many technologies exist in order to control home devices. The most popular ones include legacy technologies such as KNX or X10 and recent alternatives such as ZigBee, Z-Wave, WiFi, INSTEON, or Bluetooth Low Energy.

11.3.2 X10

X10[20] is a communications protocol designed for remote control of electrical devices using the existing electrical power line (220 or 110 V) inside the house. It was developed in 1978 by Pico Electronics of Glenrothes, Scotland, and constitutes the first domotics protocol in the market. However, it is still widely used nowadays despite its simplicity and limited functionalities. X10 uses a very low transmission rate on the order of bits per second, enough for transmitting its control commands. It may control a maximum of 256 devices, and supports six basic functions: on, off, lower, intensify, all on, all off. X10 comprises lamp modules (dimmer, etc.) and device modules (just on/off commands). There are bridges to translate X10 to other domotics standards such as KNX, and vice versa.

The X10 protocol is simple and it consists of bits representing addresses and bits for commands, for example, "lamp number 2" (address) "off" (command). A remote control (controller) can be used to wirelessly transmit the command to the transceiver module, which is then switched to the electrical power network to transport the signal to the appliance (e.g., lamp) module that executes the command, acting as an actuator.

X10 uses the zero crossing of the power to which it is wired to transmit information. This zero crossing point is the instant where there are fewer electrical disturbances, which are the main problem of the power network that causes interference, and allows for the synchronization of both the transmitter and the receiver. A 120 kHz pulse (high frequency compared to the 50 Hz sine wave of the electrical signal) at the zero crossing point means a digital "1," and no pulse means a digital "0." Binary digits are sent along with their complementary through a complete electric signal cycle, which means that a "1" is represented by a 120 kHz pulse followed by no pulse at the adjacent zero crossing, and vice versa for a "0." The start code is formed by three pulses followed by no pulse within two electrical signal cycles. A letter code is sent after the start code, and it is composed of four bits (four cycles). Following this, a second group of four bits (or nibble) corresponds to the number code, and finally the bit function to form a frame of 11 cycles. The

function bit indicates whether the letter code means a command (function bit equals
"1") or an address (function bit equals "0"). For reliability issues the frame is transmitted twice. Any address code must be separated from any command code by a
buffer of six zero crossings, or three electrical power cycles. Therefore, a standard
X10 transmission lasts 47 cycles of AC power. Address frame is avoided in all on/
off commands.

11.3.3 KNX

KNX[21] is a standard (ISO/IEC 14543) network communications protocol used in
domotics. KNX is the successor and the convergence of three previous standards[22]:
the European Home Systems Protocol (EHS), BatiBUS, and the European Installation Bus (EIB or Instabus), and is managed by the KNX Association. KNX proposes a distributed network with intelligent elements, and defines several
communication channels: twisted pair wiring (inheritance from BatiBUS and EIB
Instabus at 9600 bps), electrical power (inherited from EIB and EHS, similar to
X10, at 1200/2400 bps), radio (Industrial, Scientific and Medical (ISM) band at
16,384 kbps with low energy consumption), infrared (for remote controls), and
Ethernet, also known as EIBnet/IP or KNXnet/IP (at 10 Mbps). The local area network serves as backbone between EIB segments, and IP routers replace area and
line couplers in legacy architectures, enabling Fast Ethernet to replace the bus.

Along with the sensors and actuators, the classic KNX infrastructure consists
of gateways (routers are also feasible) to interconnect with other communication
protocols, and couplers to establish the physical separation in the bus enabling
physical addresses and dividing the network into areas, groups, and lines. However,
KNX is evolving toward an IP network that simplifies the topology and increases
the connectivity: The IP network acts as a backbone for KNX, the line couplers are
substituted by KNX/IP routers (connected to a KNX subnetworks and to the IP
backbone network), and backbone couplers would disappear.

11.3.4 ZigBee

ZigBee is the name of a group of universal and open wireless protocols based on
the wireless personal area network IEEE 802.15.4 standard. It is supported by the
ZigBee Alliance established in 1998 and constituted by over a thousand companies
in the wireless communications sector and specifically in the home automation sector. The ZigBee specification is free for noncommercial purposes. New specifications are being developed to overcome the lack of complete interoperability among
manufacturers.

ZigBee devices are small, with low complexity, low power consumption, and
low infrastructure cost, including the device itself and its set-up. A ZigBee network
can be automatically set up, its size can be scaled, and it can work with no human
intervention, and a meshed, star or tree network topology may be established, supporting up to 65,536 nodes. The ISM band at 2.4 GHz is employed defining 16 five

MHz channels. The data transfer rate is low and a maximum speed of 250 kbps can be attained using an Offset Quadrature Phase Shift (OQPSK) modulation, enough for home automation purposes. With a Binary Phase Shift Modulation (BPSK), modulation of 20 kbps can be reached in the lower part of the ISM band (868 or 915 MHz). A 1 milliwatt power limits the range to 10 m outdoors, but in turn it can operate at a low power consumption of 30 mA when on active transmission and 3 μA when idle. ZigBee employs data encryption with 128 AES to ensure security, and is able to customize security for each application.

The main ZigBee advantage lies in its flexibility, enabling the development of profiles that are specific application software that connect to the ZigBee stack and facilitate the creation of wireless products for specific applications. Among the available profiles we can find home automation, smart energy, telecommunications, health care, remote control (RF4CE, or radio frequency for consumer electronics), building automation, and retail services. The use of the same ISM band as WiFi and Bluetooth may translate into interference, minimized with coexistence mechanisms implemented in the ZigBee transceivers and also with the search of the minimum interference frequency among the 16 available channels. The limited inter-operability of ZigBee (together with the interference problems) constitutes the main ZigBee disadvantage.

11.3.4.1. ZigBee Operation

The ZigBee network follows a mixed master–slave architecture, with a coordinator node, as well as nodes acting as slaves, and other nodes acting as routers.

ZigBee devices can be classified into coordinator, router, and end device.

- The ZigBee coordinator is the most complete device and it controls the network and the paths for the devices to follow to interconnect themselves. There is one coordinator device in each network.
- The ZigBee router interconnects separated devices along the network topology.
- Finally, the ZigBee end device communicates with its parent node (being a coordinator or a router) and cannot transmit information intended for other devices. It is a simple device that can be idle most time, maximizing its battery and with minimum memory requirements, therefore being less expensive.

Depending on its functionality, a ZigBee device can be considered as an active or passive node.

- A Full Functionality Node (FDD) is an active node that receives messages in 802.15.4 format. Its additional memory and higher computing capacity enables a ZigBee FDD to act as a coordinator or a router, or to be used in network devices that act as interfaces with the users.
- A Reduced Functionality Device (RFD) is a passive node with limited capacity and functionality that achieve a low cost and extremely simple device. Sensors and actuators are RFD.

Any FFD or RFD device can decrease its power consumption due to its capacity to remain idle most of the time. A ZigBee node is able to wake up in less than 15 ms (and back to idle when not required any more), to establish a network link in less than 30 ms, and to access the channel in less than 15 ms.

The ZigBee network topology can be either star, mesh, point to point, or tree.

- In a star topology, one of the FFD devices is the network coordinator, which initializes and maintains the rest of the devices in the network. Each device communicates directly with the coordinator.

- In a mesh topology, the coordinator initializes the network and selects the parameters to use. The network may be extended by means of ZigBee routers with up to 254 nodes. With local addressing, up to 65,000 nodes can be set up.

- A tree topology is a special case of a point-to-point topology in which there is one FFD coordinator and every device may communicate with another one within coverage. In this topology many devices are FFDs, and RFDs may connect at the end of the network. A ZigBee network including multiple topologies can be established.

11.3.5 Z-Wave

Z-Wave is a proprietary wireless protocol initially developed by Zensys and bought by Sigma Designs in 2008. Since the protocol is not open, it is only available to customers of the previously mentioned companies. The physical and MAC layers of Z-Wave have been included by ITU as an option in its G.9959 standard, which defines a series of guidelines for narrowband wireless devices operating under 1 GHz. Z-Wave is supported by the Z-Wave Alliance, which has over a hundred companies in the home automation sector with hundreds of interoperable products.

11.3.5.1. Z-Wave Operation

Z-Wave, such as ZigBee, follows a master–slave architecture, but there are several differences in the operational mode and technical features. Z-Wave consists of a simple low-cost meshed network operating in the part 15 of the ISM band (868–915 MHz) attaining bit rates of 9600 bps and 40 kbps, with an irradiated power of 1 mW that covers up to 30 m. Within the meshed network any node can communicate with an adjacent node, and in case they are distant enough not to be in range, they may link to an intermediate node reachable by both of them to establish the communication. Two types of identifiers are defined in the Z-Wave protocol:

- A Home ID, with four bytes, a common identifier for every node belonging to a logical network.

- The node ID, which establishes the address of a unique node in the network. The node ID is formed by one byte, so there is a maximum of 2^8 supported devices. Taking away the ones reserved for special functions there last 232 nodes remaining to support the Z-Wave network.

The Z-Wave devices can be classified into controllers and slaves.

- A controller controls other Z-Wave devices, it is factory set and its Home ID initially cannot be changed. The main controller assigns the nodes its Home ID, which then belongs to the controller's network. In the inclusion process, the main controller assigns an individual Node ID to every new device that is added to the network.
- The slaves are devices controlled by other Z-Wave devices and they adopt the Home ID the network is assigning them.

In case there initially are two controllers, one of them is selected to be the main controller that will assign its Home ID to the rest of devices in the inclusion process, including the secondary controller, which cannot include any more additional devices. Also, individual Node IDs will be assigned to every device by the main controller.

Z-Wave implements routing to guarantee the wireless link availability between the main controller and the rest of the nodes and avoids link cutoffs. The routing process allows the nodes to resend and repeat messages to other nodes unreached by the controller, with up to four repeaters. Therefore, the more the nodes, the more flexible and robust the network. Each node is able to determine which nodes are reachable, the neighbor nodes, and informs the controller that creates a routing table with all possible routes in the network. The routing table can be partially accessed by the routing slaves, but slave devices have no information about it. In fact, a slave device can only reply to the node from which it has received the message. It cannot send unsolicited messages.

11.3.5.2. *Z-Wave versus ZigBee*

Z-Wave and ZigBee present several similarities that are also advantages, and make them both a very extended solution among manufacturers nowadays. Z-Wave and ZigBee have been in the market for many years and have become alternatives. In fact, they are the most named standards among 802.15 networks. They offer low power consumption and are based in smart meshed networks with the ability to add more devices to increase control.

The main difference between both protocols lies in the interoperability issue. ZigBee is an open protocol that each manufacturer adopts according to the standard specifications, and it guarantees interoperability at the physical and network layers but not at the application layer. Therefore, a ZigBee product from one manufacturer may not communicate with a ZigBee product from a different manufacturer. On the contrary, Z-Wave is a proprietary protocol that guarantees interoperability among all Z-Wave products, but in turn it might be affected by what becomes to the company owning the protocol. ZigBee and Z-Wave also differ in the technical features such as the number of supported nodes (a maximum of 65,536 for ZigBee versus 232 for Z-Wave), the transmission rate (up to 256 kbps for ZigBee and up to 9.6 kbps for Z-Wave), the operation frequency (ZigBee at 2.4 GHz and Z-Wave at 868–915 MHz), and the range (ZigBee with 10 m and Z-Wave with up to 30 m).

11.3.6 Bluetooth Low Energy

Bluetooth Low Energy (BLE), also known as Bluetooth Smart or Bluetooth 4.0+, is a Bluetooth low-power version that implies a longer battery life. It has the advantage of being compatible with any Bluetooth device, and with a very high market penetration. The Bluetooth 4.0+ specification covers both classic Bluetooth and BLE, although the former Bluetooth versions are not directly compatible as we will see in the following paragraphs. Nowadays many devices implement BLE (e.g., e-health sensors) adopted by big companies in the smartphone market that include Bluetooth 4.0+.[23] BLE offers interoperability among manufacturers, low implementation costs, and an increased range that varies depending on the application.

11.3.6.1. Bluetooth Low-Energy Operation

Bluetooth Low Energy is not backward compatible with Bluetooth: Bluetooth 4.0+ specification allows the devices to implement one of them, or them both. Current smartphones may support both classic and smart Bluetooth with hardware and software. Two tags are used: Bluetooth Smart Ready is employed in dual-mode devices compatible with classic and smart peripherals, and Bluetooth Smart used in single-mode (smart only) devices that need another Smart or Smart Ready device to work.

- In dual-mode chips, BLE is integrated in a classic Bluetooth controller (Bluetooth v2.1+EDR or Bluetooth v3.0+HS), therefore increasing the functionalities of the devices.
- Single-mode chips are compact and integrated devices with a light link layer that allows for ultralow-power operation, easy device discovery, and reliable point to point data transfer.
- Dual-mode chips employ the BLE part of their architecture to communicate with single-mode chips. However, single-mode chips cannot communicate with classic Bluetooth devices that do not implement Bluetooth 4.0+.

Bluetooth Low Energy operates in the 2.4 GHz ISM band and irradiates a maximum power of 10 mW, being able to reach distances of over 100 m. It has a low latency, supporting a connection and data transmission time of 3 ms. It supports very small data packets with a transmission speed of 1 Mbps, using 40 2 MHz channels instead of 79 1 MHz channels of Bluetooth. On the other hand, it employs adaptive Frequency Hop Spread Spectrum (FHSS), such as Bluetooth, together with a 24 bit CRC (Cyclic Redundancy Check) in order to minimize interference from other technologies using the 2.4 GHz ISM band. Encryption in Bluetooth LE uses AES-128-CCM cryptography. BLE includes a smart controller that allows the host to sleep for long periods of time, being awakened by the controller only when the host needs to perform any action. This implies high energy saving since the host power consumption is higher that the controller's.

Like classic Bluetooth, BLE defines profiles that specify how a device should work for a specific application. Manufacturers must implement these profiles to ensure compatibility. Current profiles are based on Generic Attribute Profile (GATT), a general specification to send and receive a small amount of data (the attributes). Examples of profiles for specific use cases include the following: for e-health—temperature measurement, glucose monitoring, and blood pressure; for sports—heart beats and running speed; others—proximity sensors, alerts, and time.

Bluetooth Smart topology may be broadcast, connection, or mixed.

- In a broadcast topology, unidirectional information is sent and received by any listening device. Roles include the broadcaster (sending periodic advertising packets) and the observer (regularly scanning its preconfigured frequencies for listening to the sending packets). Broadcast is an easy and rapid mechanism and the only alternative to send data to several peer devices simultaneously. Its main drawback is the lack of security: any listening device will receive the broadcasted data.

- The connection topology implies a private, permanent, and periodic data exchange between two devices. It is commonly employed when transmitting bidirectional data or a higher data volume than that supported by advertising. Compared to broadcast, it is energy saving. Devices adopt two roles: central (master) and peripheral (slave). The master periodically scans the preset frequencies to receive the connectable advertising packets, and conveniently starts the connection. Once connected, the master manages the communication and starts the data exchange. The slave sends connectable advertising packets and accepts incoming connections, following the timing as imposed by the master within an active connection.

In classic Bluetooth, all the slaves are listening to incoming connections, which forces them to be in a continuous standby (and then increasing the power consumption), unlike Bluetooth Smart where the master, which has less power restriction, is in charge of listening to the requests and establishing the connections.

Recent Bluetooth specifications such as 4.2. (an important Bluetooth Core Specification update) make Bluetooth Smart faster and smarter, perfectly adapting it to IoT.

11.3.7 Insteon

Insteon[24] is a network protocol for managing the digital home employing dual-mesh technology, designed by SmartLabs, Inc. Home devices are connected to the network via electrical wiring, radio frequency or both, additionally providing a backup system in case of wireless interference.

11.3.7.1. Insteon Operation

Insteon conforms to a peer-to-peer network in which devices need no supervision, therefore no controllers or routing tables are needed, and any device may act as a controller, repeater, or receiver (so devices are peers). All devices repeat the same message simultaneously, using the power line as a synchronization reference, and transmissions collide in a synchronous way preserving the message integrity. PSK modulation is employed, and Insteon products include error detection and automatic correction prior to retransmission. Insteon has the advantage of easy installation (that may be deployed in stages) since it uses the existing electrical power line combined with a wireless network and without the need of routing, it is low cost, and may be remotely controlled by a personal computer or a smartphone.

Compared to Z-Wave, Insteon theoretically increases range and reliability due to its use of the electrical wiring together with RF, the only transmission medium of Z-Wave. Also, Z-Wave employs a routed network (instead of Insteon simulcast) with a network controller and an inclusion process for each device that increases complexity. In fact, one of Insteon advantages is its ability to create a huge scaled infrastructure.

Regarding ZigBee, Insteon radio frequency (915 MHz) is not interfered by WiFi networks that employ ZigBee 2.4 GHz band, and its network is optimized for home operation. Besides, Insteon devices are peers and send messages via simulcasting, while ZigBee routes them and its devices may have full or reduced functionality.

11.3.8 Thread

Thread[25] is a protocol developed by a workgroup of several companies (with Samsung, Nest Labs – Google- or Silicon Labs among them), based on the IEEE 802.15.4 wireless protocol, currently widely used in millions of wireless devices that may implement Thread with an easy software upgrade, even some ZigBee devices could become Thread with this upgrade.

11.3.8.1. Thread Operation

Thread devices employ IPv6 and have direct Internet access, which has also drawbacks such as the need for a simple on–off switch to enter Internet in order to activate, or the obligation for the devices within the home to change their IP address if the home IP address is modified. Therefore, devices must regularly connect to the Internet, which implies some power consumption.

Thread is a protocol designed for control and automatization, so its transmission rate is low, up to 250 kbps. It operates in the 2.4 GHz band, uses a meshed network topology supporting up to 250 devices, and offers low energy consumption, low latency less than 100 ms and AES encryption.

Regarding power, thread target devices include the following:

- *Normally powered:* gateway, lighting, appliances, smart meter, garage door opener, HVAC (Heating, Ventilating, and Air Conditioning) equipment, smart plugs or fans.
- *Powered or battery:* thermostat, light switches, smoke detectors, in-home display, shades or blinds, door bell, glass break sensors, or robots and cleaners.
- *Normally battery:* door, window, motion and body sensors, door locks, or radiator valves.

Thread topology is IP-meshed network. Device-to-device communication is established within the HAN (Home Area Network) for operations in the home. A border router connects to Internet and forwards data to the cloud, and also provides WiFi connectivity to wireless terminals such as smartphones or tablets, thus reducing terminal latency within the home but connecting to the cloud when not at home for control operations. Devices join as router eligible or end device. A router eligible can become router if needed, and one of the routers (the first one to join the network) is the leader, who will make decisions on the network and may promote router eligible devices to routers to improve connectivity if required. If a leader fails, another router will become leader. End devices route through a parent device and may be in sleepy mode to save energy. A sleeping device is not required to check in to allow low power operation and polls its parent for messages, which are held by the parent.

11.4 DIGITAL HOME NETWORKS

11.4.1 Introduction

With the popularization of wireless networks at home, especially WiFi, the classic distinction between data and multimedia networks is diluted, since currently WiFi networks do not distinguish the type of data it is transmitting, whether it is general information or multimedia. However, its bandwidth requirements are higher every time, as opposed to the very narrow bandwidth of control networks. Besides, for multimedia communications there exist standards that allow for seamless connection among devices. In this section, technologies with high projection for use in a digital home are revisited. Some of them are covered in the other sections.

11.4.2 IEEE 802.11ad

During the last decade different wireless technologies grouped under the commercial name WiFi have provided higher data rates, better coverage, and other benefits that make them complement and even substitute wired networks, also offering freedom of movement to digital activities inside and outside the home or working place.

IEEE 802.11ad[26] or multigigabit employs the 60 GHz band for transmitting, enjoying a large bandwidth available worldwide (up to 9 GHz). However, the use of high frequencies involve some constraints due to the short reach of the signal (5–10 m), therefore it must be employed intraroom with direct line of sight. It may provide initial speeds of 4–7 Gbps, upgradable to 80–100 Gbps in future versions. Its uses include the following:

- Instantaneous synchronization
- Instantaneous docking, simplifying the peripherals plug-in (avoiding multiple ports and different connectors)
- Internet access complementing 802.11ac for short distance communications
- Wireless multimedia transmission of audio and video, peer-to-peer and even without compression. For example, home cinema, TV, and so on
- The first smartphone that incorporates IEEE 802.11ad is to be available soon[27]

11.4.3 WiFi Direct

WiFi Direct[28] is an ad hoc WiFi network that establishes a direct connection among two or more devices. It has no Access Point (AP), generally the router to which the devices connect to access the Internet or interconnect among them. It is a WiFi certification program that creates an AP using software, and only one of the devices needs to integrate WiFi Direct. Both Bluetooth and WiFi Direct pair several devices to data exchange allowing functions such as printing, sharing, displaying, or synchronization. However, WiFi Direct offers a larger coverage and higher transmission speeds.

An interesting application of WiFi Direct is tethering. In a tethering process, a wireless device with Internet connection acts as a gateway in order to share its Internet access to other devices, which may not have network connection (e.g., a WiFi tablet without 3G/4G) or may enjoy the device acting as router's data plan. Tethering can be established through WiFi Direct but also via Bluetooth or an USB wire. Many current smartphones implement the tethering option.

11.4.4 LiFi

LiFi[29] stands for Light Fidelity, also known as Visible Light Communications (VLC) and Optical WLAN. It is a communications technology based on light that provides high-speed bidirectional wireless communications in a similar way to WiFi. LiFi is a category of Optical Wireless Communications (OWC) that includes infrared, ultraviolet, and visible light communications. The same visible light employed for illumination is used for communications. Basically, LiFi architecture is composed of a modulator that turns on and off a light bulb very quickly, modulating the light intensity in an imperceptible way for the human eye, and a light photodiode that receives the changes in the light intensity and converts them to the electrical domain. LiFi's advantages versus WiFi include capacity, energy efficiency, and security. Main technical features include the following:

- *Bandwidth:* No licensed spectrum, which is four orders higher than that of RF, from 400 to 800 THz.
- Data density much higher than that for WiFi, since visible light can be condensed in an illuminated area while RF signal expands and might cause interference.
- Simple capacity planning, establishing light points in the communication areas.
- *Low cost:* It requires less components than RF.
- Energy efficient: LED light is already efficient and data transmission requires no additional energy.
- Security:
    - no interference with sensitive electronic circuits in certain environments
    - no listening due to light confinement (it does not cross walls)
    - data may be directed from one device to another (using a directional beam), in that case no additional security is needed as RF pairing in Bluetooth

Finally, LiFi may be used in street devices such as street lamps, bus stops, and so on. Its main drawback lies in its reduced coverage, about 10 m, and operates mainly in a direct line of sight.

11.4.5 Small Cells

The term small cells stands for low-power nodes controlled by an operator, with the aim of improving indoor coverage and cellular capacity

- may operate in the ISM band
- with an operating distance that ranges from 10 m to hundreds of meters, and according to the coverage area can be classified into
    - *Femtocell:* at home
    - *Picocell:* for business
    - *Metrocell and microcell:* rural and urban public areas
- equivalent capacity of a cellular sector
- achieve high device power saving with respect to the classic cellular network
- do not require WiFi compatible terminals

11.4.6 HomePlug

HomePlug[30] is the international standard IEEE 1901 that enables the creation of a communications network through the power line, with the use of electrical smart plugs at the power socket. One smart plug is connected to the router through Ethernet wiring, and other smart plugs provide Ethernet ports or WiFi access (providing redundancy and reliability against failure) to the devices that require Internet access in other rooms. HomePlug covers today's demand for connection by multiple devices in every room, for example, High-Definition video streaming and fast Internet access, without the need of additional wiring, extending the WiFi coverage to areas not originally reached by the wireless router and in a seamless manner to the end user.

The HomePlug standards include the following:

- *Multimedia distribution standards*
  - *HomePlug AV:* Baseline of IEEE 1901 at 500 Mbps
  - *HomePlug AV2:* It allows for next-generation multimedia networking, thanks to its gigabit speeds. AV2 technical improvements include 30–86 MHz additional bandwidth, MIMO with beamforming, efficient notching, higher order modulation and code rate, and power save modes
- *Smart energy*
  - HomePlug Green PHY for command, control, and automotive applications. It offers lower data rates of up to 10 Mbps, enough for the above mentioned purposes; it is intended for connecting smart meters, thermostats, appliances, plug-in electric vehicles. It saves up to 75% energy versus HomePlug AV.

11.4.7   IEEE 1905.1: Hybrid Home Networking Standard

The standard IEEE 1905.1[31] defines a network enabler for home networks that integrate the following wired and wireless technologies to support whole home connectivity:

- IEEE 802.11 (marketed under the WiFi trademark)
- IEEE 1901 (HomePlug, HD-PLC) power line networking
- IEEE 802.3 Ethernet
- Multimedia over Coax (MoCA)

It is also known by its commercial brand nVoy. The IEEE 1905.1 provides a communications protocol common to the nVoy-certified products, allowing users and Service Providers to take advantage of the wired and wireless technologies that already exist in the homes today. The network enabler constitutes the interface with each network. HomePlug is one of the wireless technologies integrated into IEEE 1905.1.

ACRONYMS

AP   Access Point
BLE   Bluetooth Low Energy

CRC   Cyclic Redundancy Check

DECT   Digital Enhanced Cordless Telecommunications

DLNA   Digital Living Network Alliance

DVB-T   Digital Video Broadcasting-Terrestrial

FDD   Full Functionality Node

FHSS   Frequency Hop Spread Spectrum

FTTH   Fiber-to-the-Home

GATT   Generic Attribute Profile

HAN   Home Area Network

HD   High Definition

HVAC   Heating, Ventilating, and Air Conditioning

ICT   Information and Communications Technology

IoT   Internet of Things

ISM   Industrial, Scientific and Medical

LiFi   Light Fidelity

LTE   Long-Term Evolution

MOCA   Multimedia over Coax

OWC   Optical Wireless Communications

PLC   PowerLine Communications

RFD   Reduced Functionality Device

STP   Shielded Twisted Pair

UHD   Ultra-High Definition

UPnP   Universal Plug and Play

USB   Universal Serial Bus

UTP   Unshielded Twisted Pair

VLC   Visible Light Communications

VOD   Video on Demand

NOTES

1. ZigBee Alliance, //www.zigbee.org/ (last accessed March 15, 2016).
2. Z-Wave, http://z-wave.sigmadesigns.com/ (last accessed March 15, 2016).
3. Bluetooth Low Energy, //www.bluetooth.com/Pages/Bluetooth-Smart.aspx (last accessed March 15, 2016).
4. HomeKit, //www.apple.com/ios/homekit/ (last accessed March 15, 2016).
5. The first products for Google's new smart home platform are being announced, 05/01/2016, www .theverge.com/2016/1/5/10714466/google-brillo-weave-first-products-announced-ces-2016 (last accessed January 5, 2016).

6. Samsung SmartThings, www.samsung.com/uk/smartthings/ (last accessed March 15, 2016).
7. WiFi Alliance, www.wi-fi.org/ (last accessed March 15, 2016).
8. Bluetooth, www.bluetooth.com/ (last accessed March 15, 2016).
9. DLNA, www.dlna.org/ (last accessed March 15, 2016).
10. Open Connectivity Foundation, http://openconnectivity.org/ (last accessed March 15, 2016).
11. High Efficiency Video Coding (HEVC), Fraunhofer Heinrich Hertz Institute, last access 15/03/2016, http://hevc.info/ (last accessed March 15, 2016).
12. FTTH Fiber-to-the-Home Council Americas, www.ftthcouncil.org/ (last accessed March 15, 2016).
13. 3GPP The Mobile Broadband Standard, LTE, www.3gpp.org/technologies/keywords-acronyms/98-lte (last accessed March 15, 2016).
14. Regulation of Common Telecommunications Infrastructure for access to telecommunication services inside buildings and of the activity of telecommunications equipment and systems installation, National regulation (Spain), *Reglamento de Infraestructuras Comunes de Telecomunicación para el acceso a los servicios de telecomunicación en el interior de los edificios y de la actividad de instalación de equipos y sistemas de telecomunicaciones. Real Decreto 346/2011, de 11 de marzo y Orden 1644/2011*.
15. See Endnote 13.
16. Ethernet Alliance, www.ethernetalliance.org/ (last accessed March 15, 2016).
17. H-PLC Alliance, www.hd-plc.org/ (last accessed March 15, 2016).
18. DECT Forum, www.dect.org/ (last accessed March 15, 2016).
19. Universal Serial Bus, www.usb.org/ (last accessed March 15, 2016).
20. SmartHome USA, www.smarthomeusa.com/how-x10-works/ (last accessed March 15, 2016).
21. KNX, www.knx.org (last accessed March 15, 2016).
22. CENELEC, European Committee for Electrotechnical Standardization, www.cenelec.eu/ (last accessed March 15, 2016).
23. Developers wanted: Bluetooth low energy is the future of wearables, www.broadcom.com/blog/ces/developers-wanted-bluetooth-low-energy-is-the-future-of-wearables/ (January 6, 2015).
24. Insteon, www.insteon.com/ (last accessed March 16, 2016).
25. Thread, www.threadgroup.org/ (last accessed March 16, 2016).
26. WiGig® and the future of seamless connectivity (2013), www.wi-fi.org/file/wigig-and-the-future-of-seamless-connectivity-2013 (last visited April 15, 2016).
27. First Snapdragon 820 powered smartphone announced at CES, //www.qualcomm.com/news/snapdragon/2016/01/05/first-snapdragon-820-powered-smartphone-announced-ces (last visited April 15, 2016).
28. Wi-Fi Direct. Portable Wi-Fi® that goes with you anywhere. Wi-Fi Alliance, //www.wi-fi.org/discover-wi-fi/wi-fi-direct (last visited April 15, 2016).
29. What is LiFi?, PureLiFi, http://purelifi.com/what_is_li-fi/ (last visited April 15, 2016).
30. HomePlug Alliance, //www.homeplug.com/ (last visited April 18, 2016).
31. nVoy hybrid networking, //www.nvoy.org/ (last visited April 18, 2016).

Chapter 12

Videoconference and Telework

Since a few years, advanced videoconferencing (i.e., telepresence) systems have resembled quite well the feeling of being in a remote place. With recent progress in the availability of ultrabroadband, Telcos have offered managed telepresence systems for their corporate customers, but now the frontier between telepresence and High-Definition videoconference is blurring, making it available even for residential customers. In this sense, telco Comcast partnered with over-the-top player Skype but closed the service afterward without managing to attract customers willing to pay for the service.

A key use case of videoconferencing at home is telework, or telecommuting, for which videoconference is a key enabler. Telework offers benefits for

- employees: save time and money, improve life quality
- employers: save office costs, improve productivity
- society: positive impact on environment.

However, a change of mind-set is needed, and although regulation can foster it (e.g., in United Kingdom all workers have the right to request it), cultural barriers by employers hinder its broader adoption. Many telcos offer telework services to their enterprise customers, especially in those countries where it is available for civil servants in the public sector.

12.1 CUSTOMER NEED: TELETRANSPORT

A very old customer need is the ability to be in other remote places. Transportation has been evolving to meet this need, through faster and more convenient means of transport. The oldest means of walking and running were replaced by riding on animals, and then by machines, either collectively using the ship or the train, or privately using the car. More recently appeared those based on navigating through the air, with planes becoming the fastest way of travelling around the world. But

Digital Services in the 21st Century: A Strategic and Business Perspective, First Edition.
Antonio Sánchez and Belén Carro.

even planes can take long hours to take people to distant places. And even space rockets are too slow to reach other points in the space. It has always been a dream to be able to travel instantaneously to distant places. Teletransportation, or commonly teleportation, is the name given to this dream, for which transportation technology has not been able to find a solution yet.

Information and Communication Technologies have so far provided the closest experience to teletransportation, resembling with very high similarity the audio-visual experience of being in a remote place, although the other three senses have not been emulated yet: taste, smell, and touch; but complementary technologies such as virtual reality are also progressing in these missing aspects.

12.2 VIDEOCONFERENCE

Videoconference,[1] especially visual, is being improved as can be seen in the following sections.

12.2.1 Connectivity

A basic requirement to provide high-quality videoconferencing is to have an underlying high-speed network, which supports the required high bandwidth (several megabits or higher speed per second) of video encoding. Earlier in this book, a chapter has been fully dedicated to the topic of data connectivity, so there is no need to do a deep dive here. Just as a short note, it is worth commenting that a few years ago, the connectivity requirements were only met by enterprise connections. However, there is an expansion of ultrabroadband premises. The uplink consideration is quite important, since traditionally broadband connections have higher downlink than uplink speeds, with a typical ratio of 10 to 1, that is, 10 times more speed in the downlink. However, a videoconference requires symmetrical bandwidth, and therefore the uplink speed determines the quality that can be achieved by videoconference.

The improvement of connectivity for the mass market has been astonishing in the past years. From narrowband (modem dial-up of just a few kilobits per second), there was a transition to broadband (ADSL/cable with a few megabits per second) to current ultrabroadband generation, based on fiber either directly to the premise or very close to it.

Back to the comment about uplink speed relevance, it was somehow a limiting factor in broadband, but it is also beginning to disappear.

There is a growing number of countries in which telecommunications and cable operators are already offering ultrabroadband speeds, many of them in a symmetrical manner. For example, Spain launched in August 2015 a commercial offering of 300 Mbps symmetrical (i.e., 300 Mbps both for the downlink and the uplink),[2] available for 5 million homes. And *gigabits per second* speeds are already offered in an increasing number of countries: for example, in Japan, Korea, United States, and New Zealand.[3] In the case of United States, some announcements made since late 2014 include either pilots with limited coverage or plans for more massive rollout:

- US Internet[4] 10 Gbps symmetrical
- Google Fiber[5] 1 Gbps symmetrical $70/month (It was announced that San Antonio is the largest city in which the service is going to be deployed, with 1.4 million inhabitants covered by August 2015; the service was available earlier only in Kansas City, Provo, and Austin; besides San Antonio, it was also announced that Salt Lake City, Nashville, Atlanta, Charlotte, and Raleigh-Durham; and Phoenix, San Jose and Portland were potential cities for deployment.)
- CenturyLink[6] 1 Gbps symmetrical
- Cox[7] (gigabit speeds)
- AT&T[8] (gigabit speeds, *GigaPower* in 24 cities and 15 markets)
- Comcast[9] 2 Gbps symmetrical (coverage of 1.5 million customers in Atlanta; expected coverage of 18 million homes by the end of 2015)

With all these facts, it could be easily said that connectivity is a disappearing barrier, and it is already enabling high-quality videoconferencing.

12.2.2 Quality: Resolution and Frequency

Videoconference requires equipment to capture and display the audiovisual stream. A key feature of videoconference is resolution. Typical resolutions are already high now. The resolution unit is the pixel, which is the representation of a spatial point of a picture. For true color, a pixel occupies 3 bytes, that is, it can take almost 17 million values. In the case of RGB (Red Green Blue) coding each byte represents one of the three primary colors, the pixel is the combination of the three.

On the one hand, there is 720p (HD Ready); then 1080p or Full HD; and finally, Ultra-HD with resolution of 3840×2160, equivalent to 8 megapixels, or four times more resolution than Full HD.

It has to be noted that videoconference usually lags behind television in terms of resolution, where 4000 is more common (and even 8000 supported). The main reason is that in television, only decoding is needed, but in videoconference real-time encoding is also required, and this is computing intensive.

In Table 12.1, a summary of main video resolutions commented above is shown.

Table 12.1   Video Resolution

Name	Resolution
720p (HD)	1280×720
1080p (Full HD)	1920×1080
4K (Ultra-HD)	3840×2160

Besides resolution, another quite important feature is frequency or frame rate, with more frequent picture update giving a higher sensation of seamless video movement. The typical value is 30 fps, but 60 fps is also supported in better systems.

Multiplying the pixel size, resolution, and frame rate gives the raw video bandwidth required. For example, with 3-byte pixels, Full HD resolution and 60 fps, the required bandwidth is almost 3 Gbps. However, the video can be compressed, reducing significantly the final bandwidth requirement.

12.2.3 Compression: Codecs

Old videoconferencing standard codecs are H.261 and H.263. The current one is H.264, for which royalties apply, with no upper limit.[10] This is certainly a problem for free or even freemium applications such as web browsers or even videoconference software.

In order to tackle this, Cisco unveiled Thor project in July/August 2015, hiring codec experts (both technology and patents), open sourcing the code, and contributing it to IETF (Internet Engineering Task Force) to standardize it through its NetVC workgroup. Mozilla has also contributed its technology, *Daala*.

Immediately afterward in September 2015, the Alliance for Open Media was announced,[11] founded by Amazon, Cisco, Google, Intel Corporation, Microsoft, Mozilla, and Netflix. Its objective is to provide an Ultra-HD video codec. Its intended features are interoperable and open (royalty-free), optimized for web, bandwidth scalable, low computational requirements, and real-time delivery. It joins the forces behind *Daala* (Mozilla), Thor (Cisco), and VPx (e.g., VP9 evolving to VP10, by Google) codecs.

12.2.4 Cameras

At the low end, videoconference can work with general purpose equipment (i.e., a personal computer) that might already include a webcam for video capture, on top of more typical peripherals such as a microphone and speakers, taking for granted the inclusion of a screen. It is not infrequent to see personal computers (e.g., laptops) that include an embedded webcam with a good resolution, for example, of 720p. However, there are also external webcams that can connect to the computer via external ports such as USB-Universal Serial Bus (or even HDMI (High-Definition Multimedia Interface)).

An affordable (around €100) example of high-quality webcam is Logitech C930e[12]:

- H.264 SVC encoding (on-camera video processing)
- 1080p (1920 × 1080) resolution
- 30 fps
- 90° field of view, Carl Zeiss lens, autofocus, low light

- Shutter. Pan, tilt, digital zoom (4×)
- USB port (2.0, 3.0 ready). UVC (USB video class) 1.5
- Since 2013

A mid-market camera for rooms is ConferenceCam Connect by Logitech[13] with the following features:

- Portable
- For small- and medium-sized rooms
- USB
- Virtually any videoconferencing software: Microsoft® Lync™ and Skype™, Cisco Jabber® and WebEx®, Citrix GoToMeeting®, Blue Jeans®, Google Hangouts™, Lifesize®, Vidyo®, Zoom®, and so on.
- 90-degree field of view with digital pan and tilt, mechanical tilt control, 4× digital Full HD zoom, and ZEISS® optics with autofocus.
- 360° sound, 12-ft diameter range while full duplex sound, acoustic echo, and noise-cancelling technology.
- Available worldwide in 2015 at a suggested price of $499, €499, and £449

A high-end camera is, for example, Cisco TelePresence PrecisionHD Camera 1080p 2.5×:

- 1080p
- 60 fps
- 2.5× optical zoom
- Motorized
- Dual HDMI/Camera Control and USB output

There are other high-end cameras (with HDMI port) by vendors Polycom, Life-Size, Sony (around $3000).[14]

A high-end 4K camera was announced by Panasonic in June 2015.[15] In the same month, TrueConf claimed to pioneer 4K 1-on-1 video calls (they had already showcased 4K group video conferencing in 2014), based on a beta prototype consisting of a PC with Intel i7 fourth-generation processor, video card, Point Grey camera, its server software, and a Polymedia Flipbox display (4K), and 4 Mbps bandwidth.[16] Commercial availability could be expected for 2016, alongside more powerful PCs, affordable cameras, and hardware support of H.265 and VP9 codecs.

12.2.5 Room Systems

The highest end of videoconference room systems is telepresence systems. Telepresence are immersive, resembling same room experience through real picture size, same decoration, curved desk, and so on. Traditionally, prices have been huge, but recent systems have reduced the prices (e.g., avoiding room remediation). Sample features of a recent less expensive telepresence system (Cisco TelePresence IX5000 announced in late 2014) include the following[17]:

- Three 70-in. LCD screens
- H.265 video codec
- Three video (60 fps) and two content streams
- 4K (2160p) Ultra-HD cameras (but 1920 × 1080p video) and theater-quality audio
- One-row seats up to six people, two-row 18
- Integrated lights (LED: Light Emitting Diode)
- Without the need for costly room remediation
- One-button-to-push meeting start (to start the meeting by just pressing a button, without complex configuration)
- Document camera
- 18-54 discrete microphones (microphone array–with 17,502 sound-capturing micro holes); 3-channel AAC-LD (22 kHz) spatial audio combined with 19 high-fidelity speakers (12 main drivers, 6 tweeters, 1 subwoofer), echo cancellation
- Average bandwidth utilization for three-screen video: 720p@30 fps 1.8 Mbps (H265) or 3 Mbps (H.264). 1080p@60 fps 10.8 Mbps (H.265) or 18 Mbps (H.264)
- $299,000 full-list price (might be cheaper)

In order to compare with traditional room size HD videoconference systems, here is an example with the following features (Polycom Real Presence Group series, e.g., 500 model[18]):

- 1080p with 60 fps
- Wide angle lens
- Video codecs: H.264 AVC, H.264 High Profile, H.264 SVC (but not H.265)
- Minimum bandwidth: 1080p, 60 fps from 1740 kbps; 1080p, 30 fps from 1024 Kbps; 720p, 60 fps from 832 kbps; 720p, 30 fps from 512 kbps

In last quarter of 2015, Polycom announced new hardware,[19] including an advanced hands-free phone that supports the connection of a webcam, 360-degree voice and video technology, affordable videoconference for small rooms.

12.2.6 Software

Hardware manufacturers typically complement their product portfolio with a companion software for PCs (or even tablets and smartphones), which is a more suitable way to guarantee interoperability with their room systems. Beyond video, these applications also support content sharing (e.g., a presentation) and even remote control of the remote camera (Pan-Tilt-Zoom). Two of the most prominent vendors are Polycom and Cisco.

Polycom offers Real Presence Desktop software (30-day free trial). Its features can be seen in its data sheet.[20] It supports H.264 (including SVC and High Profile). However, it only supports resolutions up to 720p/30 fps (encode and decode); in other words, it does not support higher resolutions (1080p or 4K). It supports many audio codecs, including wideband ones and their own ones (G.711 U, G.711A, G.719, G.722.1, G.722.1C, G.722, G.728, G.729, Polycom® Siren LPR, Polycom® Siren™ 14), as well as other features such as echo cancellation. For HD video, required processors are Intel Core i3 Dual Core 2.5 GHz or higher (up to 15 fps transmission, 30 fps reception) and Intel Core i5 Quad Core 2.0 GHz or higher (up to 30 fps in both directions).

Cisco offers Jabber. On the one hand, Cisco's Jabber Video for TelePresence[21] (formerly Tandberg Movi) is being discontinued. According to specifications, it supports bandwidths up to 8 Mbps. Regarding codecs, for video it supports H.264 (and H.263), and for audio (besides G.711) wideband ones, namely MPEG4 AAC-LD (48 kHz, 64 kbps), G.722.1 (24/32 kbps). Interestingly, it supports up to 1080p (as commented, 1920×1080 pixels) 30 fps encoding and decoding, for which 2 GHz Core 2 Duo processor or better is recommended. It can be downloaded for free.

On the other hand, Cisco Jabber[22] itself (available for PC, both Windows and Mac, and mobile –iOS, Android, and Blackberry). As opposed to Cisco Jabber Video for TelePresence, the client (e.g., in Windows) only supports 720p. It is more tightly integrated with Cisco's own video infrastructure.

Talking about videoconference software, of course it is a must to mention Skype.[23] After its acquisition by Microsoft, it now offers two variants, one for enterprises (Skype for Business) and one for consumer (Skype), both of which have been progressively converging (i.e., in terms of interoperability). The consumer version is standalone and supports up to 10 people on video calls. Windows desktop version supports Full HD, 1080p. There is a web version (Internet Explorer, Chrome, Safari and Firefox) called Skype for Web (Beta), which was made available worldwide by mid-2015.[24]

Skype for Business was formerly known as Microsoft Lync. As opposed to its consumer counterpart, which is free, Skype for Business has license fees, either in online mode or based on a server. It is also more integrated with the Microsoft Office suite. Online meetings support up to 250 participants with HD video.[25] HD Video supports 1920×1080 resolution,[26] with 30 fps. It also interoperates natively with consumer Skype videoconferencing users. And it also interoperates with room videoconference systems through gateways, although the latest room systems can communicate directly without the need for additional equipment.

Of course, there are many more applications, especially those provided by videoconference and unified communications vendors. Examples are Lifesize (Logitech) Softphone, and Avaya Scopia.

Finally, it is worth mentioning Mozilla Firefox Hello.[27] The Mozilla Firefox browser includes a videoconferencing application (leveraging WebRTC-embedded technology). The user just needs to send a conversation link to the other party asking to click it.

12.2.7 Cloud Video

With the emergence of Cloud technologies that transform Information Technologies (IT), it is natural to see a new videoconference category called Cloud video. In it, videoconference backends are hosted in the Cloud (Videoconference Software as a Service), and users can connect to a videoconference meeting no matter what videoconference endpoint they use.

One of the most prominent companies in this space is Blue Jeans Networks. In September 2015 it announced a new founding round[28,29] (pre-Initial Public Offering), in which it raised $76.5 million, bringing the total overall financing to $175 million. By then, the company claimed more than 1 billion videoconferencing minutes per year (compared to 2011—when the company was launched—when the entire video conferencing market only a fifth of that, just 200 million min annually). It is serving 25 million participants worldwide. With such figures it could also claim to have disrupted the video bridging hardware (Multipoint Control Unit) market that is worth a billion dollars annually. Its platform and APIs (Application Programming Interface) allow one to embed videoconference in third party applications. Beyond that, in 2015 the company released a second product to host very large-scale online events, complementing traditional broadcast with interactivity with the audience. The company has almost 5000 customers, who collectively have saved 7.5 billion travel miles, $3.3 billion in travel costs, and 2.7 billion pounds of CO_2 emission. BlueJeans supports many different endpoints:

- Web browser: In September 2015 it announced support for WebRTC,[30] which is embedded in the browser and does not need to install additional software (plug-in). BlueJeans started with Google Chrome, followed by the support of Mozilla Firefox and later Microsoft Edge. Before including WebRTC support, the web browsers already supported were Internet Explorer, Google Chrome, Firefox, and Safari.[31]
- Desktop application
- Mobile: iOS and Android
- Rooms systems: anyone based on protocols H.323 and SIP, from vendors such as Cisco (that acquired Tandberg), Polycom, and LifeSize (acquired by Logitech)
- Microsoft Lync
- Other third-party desktop applications: Cisco Jabber, Polycom RealPresence, LifeSize Softphone, and Avaya Scopia
- Google Hangouts
- Phone: dial-up is still a needed feature, since meetings typically include participants that are not able to join by video (lack of hardware or software, or poor Internet connectivity that does not support a video connection)

Datasheet including video specifications for different endpoints can be found in the website.[32] It is also worth noting its scalability, supporting up to *100 endpoints* in a meeting. Furthermore, there is also a product for large events broadcasting combined with interactivity by attendants (*Primetime*), which scales up to 3000 participants. Another interesting feature is recording. Finally, regarding pricing, apparently it is not disclosed publicly (it offers a free 14-day trial). In 2013, it introduced flat fee plans, with tiers according to the number of active users (unlimited-use site-licenses for between $10 and $50 per active user per month, based on an annual fee; apparently with a minimum number of active users).[33]

Another quite well-known Cloud video application is Google Hangouts.[34] With Hangouts, the host starts the meeting and shares a web address, to which users can connect from a web browser (similar to the way an audio bridge works, in which all users dial the same number). Video resolution is 720p. In its free version it offers multivideo conferences simultaneously, allowing up to 10 (or 15 with premium version) people to interact. It is also accessible online (as well as offline through recording) through YouTube (Live Events), enabling massive streaming.

Another relevant company is Tokbox,[35] which develops and operates a video Cloud platform based on WebRTC (which is natively supported by some web browsers, as already commented). The Application Programming Interface (API) supports the following:

- Voice
- Video
- Multiparty calls
- Text chat/messaging
- Multimedia archiving
- Security with encryption
- Audio and video streams customization
- Firewall traversal
- Audio detection
- Audio fallback (audio prioritization for poor connections)
- Screen sharing
- Moderator controls

In terms of devices, it supports web browsers (Mozilla Firefox and Google Chrome through WebRTC and Internet Explorer with a plug-in) and mobile devices (based on iOS and Android, the latter also with Firefox and Chrome).

Regarding pricing, it is based on minutes, with a minimum fee of $50 per month, including 10,000 min, that is, $0.5 cents/min. This goes down progressively: $0.475 cents up to 100,000 min, $0.45 cents up to 0.5 million min, $0.425 cents up to 1.5 million min, $0.4 cents up to 5 million min, with customized prices beyond 5 million. It has to be noted that minutes refer to the number of video streams sent by the server multiplied for the duration of a call; that is, for a 1 to 1 videoconference there are two video streams, for a multiparty call of n participants, there are n multiplied by $n - 1$ video streams (e.g., for a three-way call, there are six video streams, because each participant receives two video streams from the other two participants).

12.2.8 Robotics

An interesting development in the field of videoconference are telepresence robots[36] (self-balancing and dual kickstands) with a vertical *stick* where a tablet (iPad) is inserted. The movements can be controlled remotely (speed—slow to moderate walking speed, direction, even the height is also motorized) through a web browser or an app (iOS). Its technical specifications are as follows:

- Battery lasts for 8 h and recharges in about 2 h.
- It weighs about 7 kg (including the tablet), its height ranges from 120 to 150 cm, and its footprint is just 10 in. × 9 in.
- Indoor use, mostly flat surfaces (including carpet) as well as wheelchair ramps (5%).
- It connects through WiFi or cellular (through tablet).
- Relies on tablet for microphone, speakers, and cameras (front facing and downward facing through back camera and a mirror).
- Video based on WebRTC (also with encryption).

Its price is $2499 with an AC adapter and 1-year warranty (2 additional years for $499). In terms of accessories, charging dock costs $299, and audio kit with amplified speaker and directional microphone $99. Education Discount is 4%.

As can be gathered, the main application of such a robot is for telework, allowing remote people to navigate the office with colleagues, attend to meeting rooms. Higher education (distance learning) is also another example of use case.

12.2.9 TV

For residential users, a common device that can be used for videoconference is the TV set. The videoconference application is embedded in modern Smart TVs, but legacy TVs can be enabled through an external device. The most used application for TVs is Skype[37]:

- 1080p and three-way video on 2014 Samsung model
- Built-in camera or add-on (e.g., €99)
- Major manufacturers: Samsung, LG, Sony, Panasonic, Sharp, Vizio, Philips, Toshiba
- TVCam (all-in-one) for non-Smart TVs

12.2.10 Market Size

Market trends are toward cheaper systems, personal systems, from hardware to software and Cloud infrastructure.

The main enterprise telepresence and videoconference equipment vendors are Cisco, Polycom, ZTE, Huawei, and Avaya (others include Lifesize (Logitech), Microsoft, etc.).[38] Global market size was $3.3 billion (all-time high) in 2014 (+5% year-over-year), with $1 billion milestone in its last quarter (+24% year-over-year). Endpoint shipments were 15 million units (+39%), with software growing 44%, multipurpose rooms slightly above 20%, videophone and desktop below 10%, whereas telepresence decreased more than 10%.

In second quarter of 2015, revenues were $774 million, up 8% sequentially (which in turn had rose 11% year-over-year from $714 million). Multipurpose rooms increased 15% year-over-year, whereas immersive telepresence revenue grew 57% compared to previous quarter, which in turn had decreased 27% year-over-year; in other words, quarterly growth over a previous *weak* quarter.

12.3 TELEWORK

Telework, also known as work at home, or telecommuting, has a very strong cultural component. There are companies, and more specifically their management, that prefer that their employees work in the office. That is why governments have a strong role in fostering its adoption. In the next section, an overview of the situation in a few selected countries around the world is presented.

12.3.1 Regulation

In this section, a few examples of countries (the United States, the United Kingdom, and Spain in Europe) will be discussed.

As an example, in Spain (Europe) in 2012, telework regulation was included as part of a broader work reform,[39] in employee's bylaw, with the intention to foster new ways of working, and the flexibility of companies in the work organization, increasing job opportunities and optimizing the balance between work and personal life. As opposed to the legacy regulations of working at home, this one leverages intensively new technologies. In particular, the article related to distance working refers to work that is done mainly in the home of the employee (or any other place chosen) as opposed to commuting to the office. It stipulates the need to have a formal written agreement (either initial contract or a subsequent annex). Telework employees have the very same rights (e.g., training, promotions, unions—being associated to a given office location—health, and safety protection).

Besides national work regulation, another tool for governments is the adoption of telework for their civil servants. As an example, again for the case of Spain, there is no such adoption at national level. Back in 2007, the government was about to approve it, probably becoming one of the pioneers. It had already taken most of the required steps for formal approval, including a successful trial. However, the sudden change of the minister in charge of it froze the project completely (still today it has not progressed). But at the local level, governments have approved it. One of the first was in the region of *Castilla y León* (one of the 17 regions in Spain), where it is available for civil servants since 2011.[40] It started as a very reduced pilot (six people) in one department (*Agriculture and Livestock*) in 2007. After that, in November 2009, a regulation (*Orden 2154/2009*) was approved in order to do a wider pilot with 100 employees that took place in 2010 for 6 months. Statistics show that there were 175 teleworkers (out of 465 requests) after the first 2 years, a figure that increased slightly to 182 by mid-2014. This model was later exported to other regional governments. The decree (called *nonpresential work-day through telework*, and approved on March 17, 2011) regulates the telework for civil servants. Work is developed outside the premises of the government through the use of new communication and information technologies (ICT). In terms of number of days of telework, it establishes a *minimum of 3 days per week*. The decree text states that *telework aims to achieve a better and more modern work organization by promoting the use of new technologies and **management by objectives** and contributes to better **work and life balance**, generating higher degree of **job satisfaction***. Of course, its adoption by an employee is *voluntary*, with same rights (including same salary) and obligations. In terms of eligibility, it excludes staff that works directly with citizens (e.g., registry offices, citizen care—*attention and information*), as well as those in the health and education sectors (e.g., doctors and teachers). Also those in top management roles cannot apply for it.

Talking about telework in the public administration, an emblematic example is the United States. The law (*Telework Enhancement Act*) was passed in 2010 (*An Act: To require the head of each executive agency to establish and implement a policy under which employees shall be authorized to telework, and for other purposes*) and was implemented by mid-2011.[41,42]

As a further step, U.S. General Services Administration approved (2011) a telework policy that was defined as opt out in the media (i.e., it could be understood that by default all the employees telework; therefore, those who do not wish to telework have to explicitly ask for office space).[43] Moreover, almost every employee is eligible to telework.

Data are being collected for the next annual report to Congress in 2016, the last one was published in December 2013. The status at that time was:

1. Eligibility: the share of federal employees eligible to telework had reached 47% (1 million), an increase of 49% from 2011.

2. Teleworkers: 30% increase in the number of eligible employees who teleworked. And among all federal employees the use of this flexible work practice increased from 8 to 10%, in other words, 10% of federal employees telework.

3. Written agreements: the number of employees who had written telework agreements also rose in 2012, to 267,227 from 144,851 in 2011.

One of the flagship agencies of the U.S. government implementing telework has been the Patent and Trademark Office (USPTO), but recently there has been some controversy about it.

Another interesting country is the United Kingdom,[44] which has regulated the right to request flexible working since mid-2014. All employees have the legal right to request flexible working. As part of it, working from home is included, stating that it might be possible to do some or all of the work from home or anywhere else other than the normal place of work. Employees can apply for flexible working if they have worked continuously for the same employer for the last 26 weeks. The employer considers the request and makes a decision within 3 months. If the employer disagrees, they must write to the employee giving the business reasons for the refusal. The employee may be able to complain to an employment tribunal. Employers must consider flexible working requests in a "reasonable manner."

In terms of adoption, there were 4.2 million people who usually spent at least half their work time using their home in 2014.[45] More than a third of homeworkers are employees, while the rest are self-employed or work in the family business.

Another topic worth mentioning is the international dimension of telework, or in order words, cross-border telework. There are tax and labor implications (residence permit, legal jurisdiction, etc.) that depend on the country of the employer and the country of residence of the employee. It is out of the scope of this chapter to go in depth into the topic, but just to show a few alternatives:

- The basic model is that of a freelance (directly or through a company) working for a company in another country. No labor implications and from a tax perspective can be probably handled as an export (simplifies within the European Union(EU), and there may be intermediary firms within the EU to handle export).
- The case is similar for an employee working through a local subcontractor.
- Another particular case is a multinational company that has offices in several (many or even all) countries, the contracts are local, but organizationally the employee works for another country.
- Another alternative is to have a local company contract and that the contract is flexible in the place of performance of the work, including abroad.
- A specific case is that of universities, and particularly for doctoral thesis (PhD), where the issue of international mobility is more standardized and promoted (e.g., Doctor title with international mention, which requires a stay abroad); therefore, pre-doctoral contracts normally allow it.
- Of course, a special case is that of entrepreneurs (or in general owners of companies), of which a prominent case is the founder of Virgin, Richard Branson, who promotes telework[46]

Related to taxation, in the United States, a bill (*Multi-State Worker Tax Fairness Act of 2014*[47]), was presented to Congress to avoid double taxation of employees working at home in a different state.

12.3.2 Companies

In this section a few success stories of companies with high adoption of telework will be covered.

Sixty-five percent of employees (30,000) work at home in telecommunications operator Telus (number 2 mobile network operator in Canada, approximately number 30 in the world).[48,49] It was started round 2005 with a telework pilot in one division, and after taking good notes of lessons learnt, expanded to all major locations in Canada. The approximately 20,000 employees that work at home do it either full time or part time. Moreover, real estate costs have been decreased by $50 million annually, and the company is on track to reach its objective of decreasing carbon footprint by 25% by 2020. Like civil servants, employees who serve customers in stores cannot telecommute. Besides that, the main eligibility requirement is that employees meet their performance objectives. A key success factor is training, especially managers. Remote teams often use videoconference and might meet in person only once a year.

Another example is Repsol,[50] a multinational oil company. More than 1400 employees enjoyed telework by the end of 2014. The program started with a pilot in 2008 with 131 participants. There has been a continuous increase afterward, with 1037 in 2012 and 1222 in 2013. All employees are eligible except those whose activity is directly related with physical workplace (think for example of oil tanks). Results are excellent with a satisfaction rating of 9 out of 10 by teleworkers.

A ranking of companies by percentage of *regular* telecommuters (in other words, percentage of employees who telecommute regularly) published in 2012[51]:

1. Cisco 90%
2. Baptist Health South Florida 88%
3. Accenture 81%
4. Teach For America 80%
5. Intel 80%
6. World Wide Technology 70%
7. Pricewaterhouse Coopers 70%
8. Ultimate Software 50%
9. Perkins Coie 45%
10. American Fidelity Assurance 40%

Another ranking shows the top 100 companies with remote jobs in 2015,[52] including the following in the top positions:

1. Teletech
2. Convergys
3. Sutherland Global Services

4. Amazon
5. Kelly Services
6. Kaplan
7. First Data
8. IBM
9. SAP
10. Westat
11. UnitedHealth Group
12. Dell
13. Working Solutions
14. Intuit
15. US-Reports
16. Xerox
17. PAREXEL
18. Aetna
19. Humana
20. VMware
21. Salesforce
22. American Express
23. . . .
24. Apple

A quite interesting experience related to telework happened in February 2012 in London (Olympics year), UK, when O_2 (a mobile network operator) completely closed its headquarters during one day,[53] in which 2500 employees worked away from the office. According to the statistics released, 3000 h of commuting were saved, 25,000 tons of CO_2 emissions also saved, and one third of the people reported being more productive. They commercialize telework service, leveraging their internal experience.

In a similar way AT&T (another telecommunication operator) offers its Telework Solutions[54] as part of its portfolio of government products (triggered by previously mentioned Telework Enhancement Act of 2010).

Besides videoconferencing, other typical services related to telework are broadcasting of events (live streaming) and Information Technology collaborative tools. Regarding the former, typical features are as follows:

- Professional audiovisual quality
- Internet transmission (content delivery network)
- Satellite uplink option
- Thousands of viewers

- Accessible from web browser
- Recording for offline consumption

Regarding software collaboration tools, it is typically enterprise software, but the boundary with consumer software is blurring:

- File sharing for project collaboration (local synchronization, version control, shared links, shared notes, etc.): Box, Dropbox for Business, Microsoft OneDrive
- Coauthoring in real-time (Microsoft Office365 Cloud: Office 2013/ 6+OneDrive/Office Web Apps)
- Remote desktop including control (Microsoft Skype for Business/Google Chrome extension)
- Microsoft Yammer: enterprise social network (groups)

Other tools that complement the typical portfolio of applications of a tele-worker, which are part of internal IT systems that are being progressively automated and migrated to Cloud, enabling access from anywhere:

- Intranet portal (daily news, file repository)
- Virtual Private Network (VPN) in order to access company internal files: they are used less, given the Cloud migration and access from mobile devices (smartphones, tablets)
- Helpdesk IT support (e.g., BMC)
- Finance (travel, purchases, etc.) (e.g., SAP)
- Human Resources (holidays) (e.g., SAP SuccessFactors, Workday)
- . . .

ACRONYMS

API   Application Programming Interface
Fps   frames per second
Gbps   gigabit per second
HD   High Definition
HDMI   High-Definition Multimedia Interface
IETF   Internet Engineering Task Force
LED   light-emitting diode
RGB   red green blue
RTC   Real-Time Communications
SVC   Scalable Video Coding
USB   Universal Serial Bus
UVC   USB video class

NOTES

1. CISCO TelePresence Solutions Hardware View. Cisco Systems, //www.cisco.com/web/ANZ/cpp/refguide/hview/ipt/tele.html (October 23, 2006) (//www.engadget.com/2006/10/23/ciscos-telepresence-meeting-does-video-meetings-in-ultra-hd/)
2. *Orange lanza su nueva oferta de fibra para más de 5 millones de hogares: La oferta convergente "Canguro de Orange" podrá disfrutarse con dos velocidades de fibra simétrica: 30/30 Mbps y 300/300 Mbps*, http://acercadeorange.orange.es/UpImages/files/2318/np_nuevaofertafibrasimetrica_agosto_85f236fb46bd30dc841f7aa2a.pdf
3. Gigabit residential UFB fiber coming to Hamilton, Tauranga, Wanganui + more / New Zealand Ultra-Fast Broadband rollout 50% complete. Ultrafast Broadband New Zealand, Global Voice Media, http://ufb.org.nz/gigabit-residential-ufb-fibre-coming-to-hamilton-tauranga-wanganui-more//http://ufb.org.nz/new-zealand-ultra-fast-broadband-rollout-50-complete/ (June 11, 2014/June 19, 2015).
4. 10 Gbps at $399 per month. US Internet, http://fiber.usinternet.com/fiber-in-the-news/ http://fiber.usinternet.com/plans-and-prices/ (December 2014).
5. Everything's faster in Texas: Google Fiber is coming to San Antonio. Google, googlefiberblog.blogspot.com/2015/08/san-antonio-fiber.html / https://fiber.google.com/cities/kansascity/ (August 5, 2015).
6. CenturyLink announces Platteville gigabit network, delivering speeds up to 1 Gbps. CenturyLink, //www.centurylink.com/fiber/news/platteville-gigabit-announcement.html (August 26, 2014).
7. Cox Communications First Provider to launch gigabit Internet service to residential customers in San Diego. Cox,. http://newsroom.cox.com/2015-08-12-Cox-Communications-First-Provider-to-Launch-Gigabit-Internet-Service-to-Residential-Customers-in-San-Diego (August 12, 2015).
8. AT&T has expanded availability of service to two dozen cities. AT&T. http://blogs.att.net/consumerblog/story/a7799396 (August 31, 2015).
9. Comcast begins rollout of residential 2 gig service in Atlanta Metro Area: Gigabit Pro will be available next month to more than 1.5 million customers; the service will bring the fastest speeds to the most people in the country (Comcast leapfrogs Google Fiber with new 2 Gbps Internet service: Rollout begins in Atlanta next month and will reach 18 million American homes by the end of the year). Comcast (The Verge), http://corporate.comcast.com/news-information/news-feed/comcast-begins-rollout-of-residential-2-gig-service-in-atlanta-metro-area (April 2, 2015) (//www.theverge.com/2015/4/2/8330267/comcast-2gbps-gigabit-pro-broadband).
10. World, Meet Thor—a project to hammer out a royalty free video Codec. Cisco, http://blogs.cisco.com/collaboration/world-meet-thor-a-project-to-hammer-out-a-royalty-free-video-codec (August 11, 2015).
11. Alliance for open media established to deliver next-generation open media formats; New open standard for Ultra High Definition video will enable enhanced video playback. Alliance for Open Media, 01/09/2015. http://aomedia.org/press-release/alliance-to-deliver-next-generation-open-media-formats/ (September 1, 2015).
12. Logitech Webcam C930e. Logitech, //www.logitech.com/en-us/product/webcam-c930e-business (last accessed September 13, 2015).
13. Logitech introduces first anytime, anywhere portable videoconferencing solution. Logitech, http://news.logitech.com/press-release/corporate/logitech-introduces-first-anytime-anywhere-portable-videoconferencing-soluti (2015).
14. Lifesize HD videoconferencing, systems and accessories; Polycom Telepresence cameras, EagleEye cameras; Sony SRG300H 30x 1080p/60 HD PTZ Camera $3,099.00 U.S. List Price. Lifesize; Polycom; Sony. //www.lifesize.com/en/products/video-conferencing-systems-and-accessories, //www.polycom.com/products-services/hd-telepresence-video-conferencing/realpresence-accessories/eagleeye-cameras.html#stab3, https://pro.sony.com/bbsc/ssr/cat-videoconference/cat-videoconferencehighdefinition/product-SRG300H/ (last accessed October 25, 2015).
15. Panasonic announces AW-UE70, industry's first professional 4k integrated Ptz camera. Panasonica, http://shop.panasonic.com/about-us-latest-news-press-releases/06162015-aw-ue70-4k-ptz-camera.html (June 17, 2015).

16. TrueConf Pioneers 4K (Ultra-HD) 1-on-1 Video Calls. Telepresence Options, //www.telepresenceoptions .com/2015/06/trueconf_pioneers_4k_ultrahd_1/ (June 10, 2015).

17. Immersive TelePresence. Cisco; Cisco TelePresence IX5000 Series Data Sheet; "Less Is More" as Cisco completely reimagines flagship three-screen video conferencing technology; the most sophisticated collaboration experience on the planet/Cisco unveils TelePresence IX5000 video system with lower power and bandwidth needs; Cisco launches new telepresence system, Project Squared mobile app. Cisco / v3.co.uk; ZDNet. //www.cisco.com/c/en/us/products/collaboration-endpoints/immersive-telePresence/index.html (November 15, 2015), http://www.cisco.com/c/en/us/ products/collateral/collaboration-endpoints/ix5000-series/datasheet-c78-733257.html, http:// newsroom.cisco.com/press-release-content?type=webcontent&articleId=1534267 (October 25, 2015), http://blogs.cisco.com/collaboration/ix5000-launch, //www.v3.co.uk/v3-uk/news/2381978/ cisco-unveils-telepresence-ix5000-video-system-with-lower-power-and-bandwidth-needs (November 17, 2014), //www.zdnet.com/article/cisco-launches-new-telepresence-system-project-squared-mobile-app/

18. Polycom Real Presence Group Series, Technical Overview; Data Sheet Polycom® RealPresence® Group 500 Powerful video collaboration for group conferences in a sleek design that is simple to use. Polycom, //www.polycom.com/products-services/hd-telepresence-video-conferencing/ realpresence-room/realpresence-group-series.html#stab2, //www.polycom.com/content/dam/ polycom/common/documents/data-sheets/realpresence-group-500-ds-enus.pdf (last accessed October 25, 2015).

19. Polycom marks 25 years of industry leadership by announcing innovative solutions that will transform collaboration experiences and be the first to put people at the center of collaboration. Polycom, //www.polycom.com/company/news/press-releases/2015/20151007.html (last accessed October 7, 2015).

20. Polycom® RealPresence® Desktop for Windows®; Data Sheet; Business-grade video collaboration software. Polycom. //www.polycom.es/content/dam/polycom/common/documents/data-sheets/ realpresence-desktop-windows-ds-enus.pdf (last accessed September 27, 2015).

21. Cisco Jabber Video for TelePresence (Movi). Cisco, //www.cisco.com/c/en/us/products/ collaboration-endpoints/jabber-video-telepresence-movi/index.html, //www.cisco.com/c/en/us/ products/collateral/collaboration-endpoints/jabber-video-telepresence-movi/data_sheet_c78-628609. html (last accessed October 25, 2015).

22. Cisco Jabber. Cisco, //www.cisco.com/web/products/voice/jabber.html, //www.cisco.com/c/en/us/ products/collateral/unified-communications/jabber-windows/data_sheet_c78-704195.html (last accessed October 25, 2015).

23. Online meetings with Skype. Skype/Microsoft, //www.skype.com/en/business/, http://blogs.skype .com/2015/06/04/how-to-get-the-very-best-skype-video-quality/ (last accessed October 25, 2015).

24. Skype for Web (Beta) now available worldwide. Skype/Microsoft, http://blogs.skype.com/2015/06/ 05/skype-for-web-beta-is-now-available-to-everyone-in-the-us-and-uk/ (May 6, 2015).

25. Online meetings and teamwork made easy. Microsoft, https://products.office.com/en-us/business/ office-365-video-conferencing (last accessed October 25, 2015).

26. Skype for business client video requirements. Microsoft, https://technet.microsoft.com/en-gb/library/ jj688132.aspx (last accessed October 25, 2015).

27. As easy as saying hello: Meet Firefox Hello, the easiest way to connect for free over video with anyone, anywhere. Mozilla, //www.mozilla.org/en-US/firefox/hello/ (last accessed October 31, 2015).

28. Blue Jeans Network closes $76.5 million investment to fuel global growth and power new era of video collaboration for meetings, events, and interactive experiences; NEA leads round with participation from Accel, Battery Ventures, Glynn Capital, Norwest Venture Partners, Quadrille Capital, and Derek Jeter. Blue Jeans Networks, https://bluejeans.com/press-releases/blue-jeans-network-closes-investment-to-fuel-global-growth-power-new-era-of-video-collaboration (September 23, 2015).

29. Blue Jeans Network raises $76.5M to bring video conferencing beyond the office. TechCrunch, http://techcrunch.com/2015/09/23/blue-jeans-network-raises-76-5m-to-bring-video-conferencing-beyond-the-office/ (September 23, 2015).

30. Long live the Web (RTC). BlueJeans Network, http://bluejeans.com/blog/long-live-web-rtc-1 (September 4, 2015).
31. Blue Jeans browser access. BlueJeans Network, http://bluejeans.com/works-with/browser (last accessed October 4, 2015).
32. Datasheet. BlueJeans Network, http://bluejeans.com/sites/default/files/pdf/BJN-Datasheet.pdf (last accessed October 4, 2015).
33. Blue Jeans rolls out "right size" plans; Blue Jeans Network rolls out "all you can meet" plans driving faster adoption of video conferencing across organizations of all sizes. BlueJeans Network, http://bluejeans.com/blog/blue-jeans-rolls-out-right-size-plans, https://bluejeans.com/press-releases/all-you-can-meet-plans, //www.marketwired.com/press-release/blue-jeans-network-rolls-out-all-you-can-meet-plans-driving-faster-adoption-video-conferencing-1787431.htm (May 6–7, 2013).
34. Google hangouts. Google, https://hangouts.google.com/, //www.google.com/work/apps/business/products/hangouts/, https://support.google.com/hangouts/answer/3367675?hl=en, https://productforums.google.com/forum/#!topic/hangouts/yM1kI5-B1Jo, //www.youtube.com/my_live_events (last accessed October 25, 2015).
35. Tokbox: Bring the Web to life: OpenTok is the leading WebRTC platform for embedding live video, voice, and messaging into your websites and mobile apps. Tokbox. https://tokbox.com/ https://tokbox.com/about, https://tokbox.com/pricing, https://support.tokbox.com/hc/en-us/articles/204605054-How-do-I-estimate-my-OpenTok-monthly-usage- (last accessed October 31, 2015).
36. Double robotics: work from anywhere: double gives you a physical presence at work or school when you can't be there in-person. Double Robotics, //www.doublerobotics.com/, http://www.doublerobotics.com/pricing.html, //www.doublerobotics.com/faq.html, //support.doublerobotics.com/customer/en/portal/articles/1379820-double-specifications (last accessed October 31, 2015).
37. Skype on TV—stay in touch from the comfort of your sofa / Get together on a free group video call – from the comfort of your sofa. Skype-Microsoft, //www.skype.com/es/download-skype/skype-for-tv/, http://blogs.skype.com/2015/01/16/get-together-on-a-free-group-video-call-from-the-comfort-of-your-sofa/ (October 31, 2015).
38. IHS-Infonetics Research. Multi-purpose rooms, Immersive Telepresence shine in videoconferencing market/videoconferencing equipment market bright spot in Q1: Enterprise video endpoints/videoconferencing equipment market hits $1 billion in the fourth quarter, software up 44 percent/videoconferencing equipment vendors feeling the impact of shifting market trends. //www.infonetics.com/pr/2015/2Q15-Enterprise-Telepresence-Highlights.asp, //www.infonetics.com/pr/2015/1Q15-Enterprise-Telepresence-Highlights.asp, //www.infonetics.com/pr/2015/4Q14-Enterprise-Telepresence-and-Video-Conferencing-Market-Highlights.asp, //www.infonetics.com/pr/2014/3Q14-Enterprise-Telepresence-and-Video-Conferencing-Market-Highlights.asp (last accessed October 31, 2015).
39. *Real Decreto-ley 3/2012, de 10 de febrero, de medidas urgentes para la reforma del mercado laboral.* Spanish Government/Official Bulletin, February 10, 2012, //www.boe.es/diario_boe/txt.php?id=BOE-A-2012-2076
40. *Jornada de trabajo no presencial mediante teletrabajo: Decreto 9/2011, de 17 de marzo, por el que se regula la jornada de trabajo no presencial mediante teletrabajo en la Administración de la Comunidad de Castilla y León; La Consejería de Agricultura y Ganadería inicia una experiencia piloto de teletrabajo; JORNADA DE TRABAJO NO PRESENCIAL MEDIANTE TELETRABAJO: Solicitudes y autorizaciones a 31 de mayo de 2013. Junta de Castilla y León;* //www.empleopublico.jcyl.es/web/jcyl/EmpleoPublico/es/Plantilla100Detalle/1246947697738/_/1284166231051/Comunicacion?plantillaObligatoria=PlantillaContenidoNoticiaHome (March 17, 2011), //www.jcyl.es/web/jcyl/AgriculturaGanaderia/es/Plantilla100DetalleFeed/1246464862173/Noticia/1180952766453/Comunicacion (June 6, 2007), //www.comunicacion.jcyl.es/web/jcyl/binarios/775/61/20130701_CHAC_NP%20Tabla%20usuario%20teletrabajo%20Junta%20de%20Castilla%20y%20Le%C3%B3n.doc?blobheader=application%2Fmsword&blobheadername1=Cache-Control&blobheadername2=Expires&blobheadername3=Site&blobheadervalue1=no-store%2Cno-cache%2Cmust-revalidate&blobheadervalue2=0&blobheadervalue3=Portal_SalaPrensa&blobnocache=true (March 31, 2013).

41. Telework Enhancement Act of 2010 US Government, //www.telework.gov/guidance-legislation/telework-legislation/telework-enhancement-act/ (January 5, 2011).
42. Annual report to Congress; 2013 Status of Telework in the Federal Government Report to Congress. U.S. Office of Personnel Management (OPM), December 2013. //www.telework.gov/reports-studies/reports-to-congress/annual-reports/, //www.opm.gov/blogs/Director/2013-status-of-telework-in-the-federal-government-report-to-congress/
43. New GSA mobility and telework policy, a model for federal agencies. U.S. General Services Administration, http://gsablogs.gsa.gov/gsablog/2011/11/01/new-gsa-mobility-and-telework-policy-a-model-for-federal-agencies/, http://www.gsa.gov/graphics/staffoffices/GSAteleworkpolicy.pdf (November 1, 2011).
44. Flexible working. UK Government, //www.gov.uk/flexible-working/overview (last accessed November 8, 2015).
45. Characteristics of Home Workers, 2014. Office for National Statistics, UK Government, //www.ons.gov.uk/ons/rel/lmac/characteristics-of-home-workers/2014/rpt-home-workers.html (June 4, 2014).
46. Give people the freedom of where to work. Virgin. 2013, //www.virgin.com/richard-branson/give-people-the-freedom-of-where-to-work
47. H.R.4085—Multi-State Worker Tax Fairness Act of 2014: Prohibits a state from imposing an income tax on the compensation of a nonresident individual for any period in which such individual is not physically present in or working in such state or from deeming such nonresident individual to be present in or working in such state on the grounds that: (1) such individual is present at or working at home for convenience, or (2) such individual's work at home fails any convenience of the employer test or any similar test. United States Congress, //www.congress.gov/bill/113th-congress/house-bill/4085 (March 20, 2014).
48. Case study: can a work-at-home policy hurt morale? *Harvard Business Review.* April 2015. https://hbr.org/2015/01/case-study-can-a-work-at-home-policy-hurt-morale, http://about.telus.com/community/english/news_centre/news_releases/blog/2014/09/15/new-study-demonstrates-why-canadian-businesses-should-embrace-a-flexible-working-program, http://business.telus.com/en/campaigns/telus-work-styles
49. New study demonstrates why Canadian businesses should embrace a flexible working program; TELUS Work Styles; Canada's cultural shift towards flexible work. Telus, September 15, 2014, http://about.telus.com/community/english/news_centre/news_releases/blog/2014/09/15/new-study-demonstrates-why-canadian-businesses-should-embrace-a-flexible-working-program, http://business.telus.com/en/campaigns/telus-work-styles, http://forum.telus.com/ttb/post/49739/all/canada%2525E2%252580%252599s-cultural-shift-towards-flexible-work (last accessed November 8, 2015; December 2, 2014).
50. Telework programme. Repsol, //www.repsol.com/es_es/corporacion/responsabilidad-corporativa/ante-quien-respondemos/equipo-repsol/diversidad-conciliacion/programas-para-llevarlo-a-cabo/teletrabajo/default.aspx (last accessed November 8, 2015).
51. 100 Best Companies to Work for 2012: best benefits: telecommuting. *Fortune,* http://archive.fortune.com/magazines/fortune/best-companies/2012/benefits/telecommuting.html (February 6, 2012)
52. 100 top companies with remote jobs in 2015. Flexjobs, //www.flexjobs.com/blog/post/100-top-companies-with-remote-jobs-in-2015/ (January 20, 2015).
53. Making flexible working work for you. O2, February 2012. //www.o2.co.uk/business/business-solutions/flexible-working
54. AT&T Telework Solutions. AT&T, //www.corp.att.com/gov/solution/integrated_solutions/telework.html (last accessed November 14, 2015).

Index

Digital Services in the 21st Century: A Strategic and Business Perspective, First Edition.
Antonio Sánchez and Belén Carro.
© 2017 by The Institute of Electrical and Electronics Engineers, Inc. Published 2017 by John Wiley & Sons, Inc.

205

THE COMSOC GUIDES TO
COMMUNICATIONS TECHNOLOGIES

Nim K. Cheung, *Series Editor*
Richard Lau, *Associate Editor*

The ComSoc Guide to Next Generation Optical Transport: SDH/SONET/OTN
Huub van Helvoort

The ComSoc Guide to Managing Telecommunications Projects
Celia Desmond

WiMAX Technology and Network Evolution
Kamran Etemad and Ming-Yee Lai

An Introduction to Network Modeling and Simulation for the Practicing Engineer
Jack Burbank, William Kasch, and Jon Ward

The ComSoc Guide to Passive Optical Networks: Enhancing the Last Mile Access
Stephen Weinstein, Yuanqiu Luo, and Ting Wang

Digital Terrestrial Television Broadcasting: Technology and System
Jian Song, Zhixing Yang, and Jun Wang

TV White Space: The First Step Towards Better Utilization of Frequency Spectrum
Ser Wah Oh, Yugang Ma, Edward Peh, and Ming-Hung Tao

Digital Services in the 21st Century: A Strategic and Business Perspective
Antonio Sanchez and Belen Carr